Lecture Notes in Engineering

Edited by C. A. Brebbia and S. A. Orszag

12

G. Walker
J. R. Senft

Free Piston Stirling Engines

Springer-Verlag
Berlin Heidelberg New York Tokyo

Authors
Graham Walker, Ph. D.
Dept. of Mechanical Engineering
University of Calgary
Calgary, Alberta
Canada T2N 1N4

J. R. Senft, Ph. D.
Dept. of Mathematics Computer Science
University of Wisconsin
River Falls
Wisconsin 54022
USA

ISBN-13: 978-3-540-15495-2 e-ISBN-13: 978-3-642-82526-2
DOI: 10.1007/978-3-642-82526-2

Softcover reprint of the hardcover 1st edition 1985

2161/3020-543210

This book is dedicated to
Carol and William Beale
and all the staff of Sunpower Inc., Athens, Ohio

PREFACE - PART A

This volume on free-piston Stirling engines is complementary to a major work 'Stirling Engines'* published in 1980, as a comprehensive survey of regenerative machines used for power, the conversion of heat to work. A companion volume 'Cryocoolers'** was published later. While embracing the whole field of refrigeration applications of regenerative machines it is particularly directed to low temperatures, the regimen characterized as the 'cryogenic range' restricted to temperatures less than 120 K.

In this volume we focus attention on the more limited field of free-piston Stirling engines. These are relatively recent developments where, it has to be said, the promise of the future greatly exceeds accomplishments of the past. However, the small group of dedicated folk working in this field have now put in place sufficient foundation to justify a specialist monograph on the technology. Hopefully its publication will stimulate and guide young, eager minds to the greater days that surely lie ahead. Its preparation will have been worthwhile if, a decade hence, subsequent developments have made the book obsolete, consigned to the historical section of the library.

Completion of the work has been made possible with the assistance of many others. First I have to thank my good friend and fellow author Dr. James Senft, of the University of Wisconsin, River Falls, Wisconsin. Senft wrote several chapters, contributed to others and, by his interest and enthusiasm, sustained my efforts thereby contributing much to completion of this book. He also prepared the index, a last-minute duty for which I am particularly grateful.

In many ways my work was largely that of a Levantine trader. Recognizing the likely inadequacy of my interpretation of their life's work, I invited several specialists in the Stirling engine field to contribute to this accumulation of free-piston Stirling engine wisdom. They responded magnanimously and so assured the book a substance and quality that, alone, I could not have given.

The late 1950's/early 1960's was the time for invention and re-invention of free-piston Stirling engine systems. It was stimulated by the research programme on Stirling engines, started in 1938 by the Philips Company in Eindhoven, Netherlands. By the late 1950's Philips' work had become sufficiently well known for keen minds

---

* Stirling Engines - Oxford University Press, 1980.

** Cryocoolers (2 Vols.) - Intl. Monographs on Cryogenics, Plenum Press, New York, 1983.

everywhere to speculate on alternative systems.

The free-piston Stirling engine appears to have been spontaneously conceived with some variations, by half a dozen people in various parts of the world. Most concepts were simply paper engines that died early with the well-meaning but entirely inadequate efforts of their inventors to reduce them to practice. The free-piston Stirling engine is a paradox of seeming mechanical simplicity which is, in fact, difficult and complicated to execute.

William Beale was a notable exception. Working as a professor at Ohio University at Athens, Beale conceived the free-piston Stirling engine arrangement that now bears his name. His early engines did not work and furthermore displayed no inclination to do so. With unbelieveable persistance and, as he now says, 'dogged stupidity', Beale continued his efforts, and, eventually, was rewarded with a self-sustaining engine stable in operation. Since those pioneer days he has devoted his life and effort to the development of the free-piston Stirling engine contributing as much as anyone to put the technology on an established foundation. In all this Carol Beale has supported him in a way that exceeds customary expectations of wifely duties. In recognition of this magnificent and sustained effort the book is dedicated to both of them and the staff of their company, Sunpower, Athens, Ohio.

Chapter 4 reviews the Sunpower free-piston Stirling engines and the techniques developed at Sunpower for the design and computer simulation of free-piston Stirling engines. It was prepared by Beale and his staff at Sunpower and by the co-author, Senft, a former employee at Sunpower.

Mr. E.H. Cooke-Yarborough of the Atomic Energy Research Establishment, Harwell, England, contributed Chapter 6 dealing with the Harwell Thermo-Mechanical Generator, (TMG). Although included here with free-piston engines the Harwell machine is actually a diaphragm engine of unique concept and form. It was invented by Cooke-Yarborough in the 1960's as a long-lived radioisotope power generator with substantially higher conversion efficiency than the thermoelectric systems then currently in use. The generator was a success technically, but the Strontium 90 fuel was too expensive for the potential users, the international lighthouse authorities and maritime agencies. When this became apparent the engine was converted to operate on liquid petroleum gas (propane/butane). At this point bureaucracy intervened. The engine running on propane had nothing to do with nuclear energy and so could not further be supported with Harwell funds, all specifically dedicated to nuclear research. Despite the sizeable amounts already spent on system development for commercial introduction to fill a widely recognized need there was little further work done following conversion to propane fuels. Only miniscule funds were available from such unlikely sources as the Irish Lighthouse Authority. However, with a persistence equal to Beale's, Cook-Yarborough stayed the course and recent developments indicate the unit will soon be in production.

Another long-time worker in the field, Dr. William Martini of Richland, Washington, contributed Chapter 7 dealing with miniature free-piston Stirling engines. These have been the subject of intense research and development effort for the artificial hearts sought by the National Institute of Health in Washington, D.C. For over a decade two highly competent teams of engineering researchers have laboured to develop small engines for converting radioisotope/thermal storage energy to the hydraulic or pneumatic work necessary to drive a blood pump. Martini led one of these teams and now, while consulting in the broader field of Stirling engines, maintains a close personal interest in the achievements of both groups.

Dr. Colin West of the Oak Ridge National Laboratory contributed Chapter 10, on liquid piston Stirling engines, a very special form of free-piston Stirling engine with great promise for the future. West invented the 'Fluidyne' liquid piston Stirling engine in the 1960's while working as a colleague of Mr. Cooke-Yarborough. He has recently summarized his experience in the field with an excellent monograph* which greatly expands the material he presents in the chapter here.

I am most grateful to all these contributors for the time and effort they have invested to produce a distillation of their wisdom and experience.

I also wish to acknowledge the assistance given to me in the preparation of the figures and photographs by Mr. Burt Unterburger and his charming assistants in the Faculty of Engineering at the University of Calgary.

My special thanks go to my secretarial and research assistant, Mrs. Karen Undseth without whose sustained interest and effort very little would have been accomplished.

I carried out much of the work at the University of Calgary during my normal activities as a Professor of Mechanical Engineering. I am grateful to the University for assistance in countless ways. Thanks are due to the Head of the Department, Dr. Peter Glockner, for his interest and encouragement and to my colleagues who tolerate my preoccupation with affairs that I suspect are not their principal interest. The University Grants Committee assisted with secretarial and research expenses using funds provided by the National Sciences and Engineering Research Council of Canada.

The draft was completed in the course of a Visiting Research Fellowship at the Stirling Engine Test Facility at the Royal Naval Engineering College, Manadon, Plymouth, England. I have to thank all my friends and colleagues there for the rather wet but very warm welcome they gave me and for the facilities extended to me there.

---

*Liquid Piston Stirling Engines - C. West, Van Nostrand-Reinhold Co., 1982.

My thanks and apologies go also to my students and to my children, Josephine and Christopher, who could, justifiably, have expected more of my time and attention.

Finally, I owe my greatest thanks to my wife Ann for her support and encouragement that far surpasses my reasonable due.

G. Walker
Calgary, Alberta

PREFACE - PART B

My first wish is to express my deep appreciation to Graham Walker for inviting me to join him in making this book possible. It has not only provided me with a unique opportunity to organize, clarify, and extend my own understanding of the subject, but has also served to motivate new and fruitful lines of research. Indeed, this is the hope I have for all of the users of this book: that in some chapter on some page each may find or be inspired to some idea of special personal interest or significance. I join Graham in profoundly thanking our distinguished contributors for their valuable efforts to realize this hope.

I am also grateful to my co-author for the friendship that working together has brought about. I am by far not the only one to find a friend while working with Graham Walker, and now I well understand why.

To many others I am also grateful for making my contributions to this book possible:

- To Gloria my constant companion who has so faithfully supported this and all my endeavors.

- To Victor who is becoming as skillful as his mother at resuscitating my enthusiasm.

- To all those colleagues, students and friends who ever offered me an encouraging word, especially to my department chairman, Dr. P. Vadlamudi.

- To Sherry Reis and Karen Undseth for expertly preparing the typescripts.

J.R. Senft
River Falls, Wisconsin

LIST OF CONTENTS

CHAPTER 1 - INTRODUCTION (G. Walker) 1

Definition and Nomenclature 1

History 2

Thermodynamics of the Stirling Cycle 3

The Stirling Cycle Refrigerator 5

The Stirling Cycle Heat Pump 6

The Stirling Cycle Pressure Generator 7

Practical Stirling Engines 7

Mechanical Arrangements 12

Piston-Displacer in the Same Cylinder 13

Pistons 13

Displacers 14

Advantages of the Piston and Displacer System 14

Piston and Displacer in Separate Cylinders 15

Single-Acting Engines 16

Double-Acting Engines 17

Reciprocator Drives: Kinematic or Free-Piston 18

Kinematic Drive Engines 18

Free-Piston Engines 19

Advantages of Free-Piston Stirling Engines 20

Disadvantages of Free-Piston Stirling Engines 21

Closure 22

References 22

CHAPTER 2 - FREE-PISTON STIRLING ENGINES (G. Walker) 23

<u>Part I: Elementary Aspects</u> 23

How a Free-Piston Stirling Engine Works 23

Beale Free-Piston Stirling Demonstrator Engine 27

Types and Classifications of Free-Piston Stirling Engines 29

Single-Acting Free-Piston Stirling Engines 30

i) Two-Piston Arrangements 31

ii) Piston-Displacer Arrangements 32

iii) Split-Stirling Systems 35

Double-Acting Free-Piston Stirling Engines 38

Hybrid Stirling Engines 41

Pendulum Free-Piston Stirling Engines 45

Diaphragm Free-Piston Stirling Engines 47
Duplex Free-Piston Stirling Engines 49
Part II: Dynamics of Free-Piston Stirling Engines 51
Introduction 51
Vibrating Systems: An Elementary Review 51
Degrees of Freedom 51
Free Vibration 52
Simple Harmonic Motion 53
Viscous Damping 53
Equation of Motion 54
Non-Viscous Damping 57
Systems with More Than One Degree of Freedom 58
Non-Linear Systems 62
Vector Representation of Vibrating Systems 62
Introduction 62
Vector Addition and Subtraction 64
Force Representation 64
Vector Force Polygon 65
Work Input and Power Consumption 67
Free-Piston Stirling Engines as Vibrating Systems 68
Vector Representation of Free-Piston Stirling Engines 70
a) Piston-Displacer System, Both Elements Sprung to Ground 70
b) Piston-Displacer System with the Displacer Sprung to Ground 77
c) Piston-Displacer System with the Displacer Sprung to the Piston 78
d) Split-Stirling Engine 79
e) Two-Piston Stirling Engine 79
f) Free-Cylinder Stirling Engine 81
g) Double-Acting Stirling Engine 82
Part III: Some Practical Considerations 84
Introduction 84
Piston Centering 84
Seals 87
Bearings 90
Hydrodynamic Bearings 90
Hydrostatic Bearings 91
Materials 94
Springs 96
Spring Stiffness 96
Closure 97
References 98

CHAPTER 3 - THEORETICAL ANALYSIS AND DESIGN (J. Senft) 100
Introduction 100
Elementary Design Tools 101
Ideal Cycle 101
Beale Number 102
Frequency Estimation 104

Efficiency 105
First Order Analysis Methods 106
Schmidt Analysis 106
Free-Piston Engine Dynamics 109
Linearization of Pressure 110
Second Order Analysis 112
Basic Power Calculation 113
Loss Mechanisms 115
Available Programs 121
Third Order Analysis 121
Analytic Design of Free-Piston Engines 123
Nomenclature 125
References 125

CHAPTER 4 - THE SUNPOWER ENGINES (J. Senft with W. Beale) 128
Introduction 128
Chronology 128
Sunpower Advice to the Designer of Free-Piston Stirling Engines 141
Free-Piston Engines and Rotary Motion 143
Guidelines for Free-Piston Stirling Engine Application 143
References 144

CHAPTER 5 - HYBRID OR RINGBOM-STIRLING ENGINES (J. Senft) 145
Introduction 145
The Hybrid Concept 145
The Ringbom Engine 145
Overdriven Mode Operation 147
Comparison with Free-Piston Engines 147
Basic Approaches to Design of Hybrids 147
The Overdriven Mode Operation Concept 148
Stable Operation 149
First Order Analysis 150
Isothermal Model for Hybrid Ringbom-Stirling Engines 151
A Criterion for Overdriven Mode Operation 154
Overdriven Mode Theorem 155
Application of the Theorem 155
Some Practical Considerations 156
Displacer Drive Loss 156
Top End Tuning 158
General Design Guidelines 158
Practical Ringbom-Stirling Engines 159
Nomenclature 164
References 165

CHAPTER 6 - DIAPHRAGM STIRLING ENGINES: THE HARWELL THERMO-MECHANICAL GENERATOR (E.H. Cooke-Yarborough) 166
Introduction 166
Diaphragms 168

	Displacer Design	173
	The Alternator	176
	System Dynamics	180
	Losses	182
	Viscous Losses	183
	Regenerator Losses	184
	Imperfect Heat Conduction	184
	Computed Losses	185
	Increasing the Generator Power	185
	Operating Experience	187
	Conclusions	193
	Nomenclature	194
	References	194
CHAPTER 7 -	SMALL FREE-PISTON STIRLING ENGINE: POWER SYSTEMS FOR THE ARTIFICIAL HEART (W. Martini)	197
	Introduction	197
	The Artificial Heart	198
	Stirling Engines in Artificial Hearts	198
	The Aerojet - General Engine	199
	Advantages	204
	Disadvantages	205
	Other Applications	205
	The JCGS Engine	206
	Potential Applications	209
	Comparison of the JCGS and Aerojet General Engines	210
	Thermo-Electron Free-Piston Stirling Steam Engine	211
	Conclusion	214
	References	215
CHAPTER 8 -	LARGE FREE-PISTON STIRLING ENGINES (G. Walker)	216
	Introduction	216
	Feasibility of Large Stirling Engines	216
	Applications of Large Stirling Engines	216
	Power Systems	216
	Large Cryocoolers, Refrigerating Machines and Heat Pumps	217
	The Gas-Fired Natural Gas Liquefier	219
	Conclusion	221
	References	221
CHAPTER 9 -	FREE-PISTON STIRLING CRYOCOOLERS (G. Walker)	222
	Introduction	222
	Free-Piston Stirling Cryocooler	222
	Integral Free-Piston Stirling Cryocooler	223
	Split-Stirling Cryocooler	227
	Large Stirling Cryocooler	231
	Liquid-Piston Stirling Cryocooler	232
	Conclusion	233
	References	233

CHAPTER 10 - LIQUID PISTON STIRLING ENGINES (C. West) 235
Basic Principles 236
Tuning of Liquid Columns 236
Feedback Systems 239
Pumping Configurations 245
Wet and Dry Machines 247
Flow Effects 249
Transient Heat Transfer Loss 251
Results 252
Engines for Beginners 253
Future Development 255
Nomenclature 258
References 258

SUBJECT INDEX 262

NAME INDEX 267

CHAPTER 1 - INTRODUCTION

DEFINITION AND NOMENCLATURE

A Stirling engine is a mechanical device which operates on a *closed* regenerative thermodynamic cycle with cyclic compression and expansion of the working fluid at different temperature levels. The flow of working fluid is controlled only by the internal volume changes, there are no valves and, overall, there is a net conversion of heat to work or vice-versa.

This generalized definition embraces a large family of machines with different functions, characteristics and configurations. It includes both rotary and reciprocating systems utilizing mechanisms of varying complexity. It covers machines capable of operating as a prime mover or power system converting heat supplied at high temperature to output work and waste heat at a lower temperature. It also covers work-consuming machines used as refrigerating systems and heat pumps abstracting heat from a low temperature source and delivering this plus the heat equivalent of the work consumed to a higher temperature. Finally it covers work-consuming devices used as pressure-generators compressing a fluid from a low pressure to a higher pressure.

Very similar machines exist which operate on an *open* regenerative cycle where the flow of working fluid is controlled by valves. For convenience these may be called Ericsson engines but unfortunately the distinction is not widely established and regenerative machines of both types are frequently called 'Stirling engines'.

Stirling engines are also known as hot air engines, hot gas engines and by a large number of designations reserved for particular arrangements of the engine elements. There is also an ideal thermodynamic cycle called the Stirling cycle. This is useful for an elementary description of regenerative machine operation but practical Stirling engines operate on a much complicated cycle and should never be described as 'Stirling cycle engines'.

The preferred generic title for all these different classes of machines is 'regenerative thermal machine' but this is cumbersome and, furthermore, it is almost certainly too late for anything other than the descriptor 'Stirling engine' to be widely and indiscriminately used. However because they have radically different characteristics a clear distinction should always be made between machines where the flow is controlled by:

a) valves (Ericsson engines),
b) internal volume changes (Stirling engines).

The term 'Stirling engine' as the generic title for closed-form regenerative engines originated in the early 1950's at the Research Laboratories of Philips at Eindhoven. Until then it was customary to call them hot air engines prefaced with the name of the inventor or manufacturer. The change of working fluid from air to hydrogen or helium at Philips in the early fifties made the term hot air engine inappropriate. The alternative 'Philips hot gas engine' proved less attractive than 'Stirling engine' and the latter has passed into general usage.

HISTORY

Stirling engines have a long history dating from the late 1600's but reached full flower in the 1800's when many thousands were built and used for all manner of pumping and similar work-consuming applications. They were as common as the contemporary reciprocating steam engines with the singular advantage that they did not explode as steam boilers were prone to do with depressing regularity.

Robert Stirling (1790-1878) for whom the engine is named was a minister in the Church of Scotland and for many years the pastor of Galston Parish Church in Ayrshire, Scotland. In 1815 he conceived the thermal regenerator and the original Stirling engine incorporating a regenerator. Working with his brother James, an engineer, Stirling made further substantial constributions to the technology until about 1845 but, so far as is known, was never extensively involved in their commercial application. A measure of his perceptive genius can be discerned from the fact that it was not until in 1850, following the introduction of Joule's concept of the mechanical equivalent of heat, that Rankine was able to explain satisfactorily the operation of a Stirling engine.

The 19th Century was a time of prolific invention and many different forms of hot air engines were invented, developed and marketed in Europe and North America. They were widely available and regularly included in the department store catalogues of the early 1900's for purchase 'off-the-shelf'.

The internal combustion engine, invented in the mid-1800's and, later, the electric motor gradually superseded both Stirling and steam engines by the First World War so that today Stirling engines are virtually unknown to the general public and even to most engineers.

The *renaissance* of the Stirling engine began in 1938 with the start of a research programme at Philips that was to endure for over 40 years and establish the secure foundation on which the modern technology of Stirling engines rests today. Philips have recently discontinued their efforts but the future for Stirling engines appears brighter than ever before, largely due to rapid increases in the price and the restricted availability of liquid fossil fuels for gasoline and diesel engines. The Stirling engine is omnivorous, capable of using any kind of heat source, and for this reason is emerging as the preferred coal-burning alternative to internal combustion engines for locomotives, marine propulsion, stationary power and large off-highway vehicles used for mining, forestry, agriculture and construction.

Further details of the history and contemporary development of Stirling engines were given by Walker (1980)*.

THERMODYNAMICS OF THE STIRLING CYCLE

The Stirling cycle proper is an ideal thermodynamic cycle made up of two isothermal (constant temperature) and two isometric (constant volume) regenerative processes as shown on the pressure-volume (P-V) and temperature entropy (T-S) planes in Figure 1.1.**

To illustrate the sequence of operations consider a cylinder containing two opposed pistons with a regenerator between the pistons. The regenerator may be thought of as a thermodynamic sponge, alternately releasing and absorbing heat. It is a matrix of finely-divided metal in the form of wires or strips. One of the two volumes between the regenerator and the pistons is called the *expansion space*, and is maintained at a high temperature T_{max}. The other volume is called the *compression space*, and is maintained at a low temperature T_{min}. There is, therefore, a temperature gradient ($T_{max} - T_{min}$) across the transverse faces of the regenerator, but it is assumed that there is no thermal conduction in the longitudinal direction. It is further assumed that the pistons move without friction or leakage loss of the working fluid enclosed between them.

To start the cycle, assume the compression-space piston at the outer dead point, and the expansion-space piston at the inner dead point, close to the face of the regenerator. All the working fluid is then in the cold compression space and the volume is a maximum, so the pressure and temperature are minimum values, represented by 1 on the P-V and T-S diagrams. During compression (process 1-2), the compression piston moves towards the inner dead point, and the expansion-space piston remains stationary. The working fluid is compressed in the compression space, and the pressure increases. The temperature is maintained constant because heat, Q_c, is abstracted from the compression-space cyclinder to the surrounds.

In the transfer process 2-3, both pistons move simultaneously, the compression piston towards (and the expansion piston away

---

* Complete details of these references are contained in the bibliography at the end of the chapter.

** Areas on the P-V and T-S planes indicate work done and heat transfer respectively. Readers unfamiliar with this representation will find it discussed more fully in my book 'Stirling Engines', 1980, OUP or in any elementary thermodynamics text. G.W.

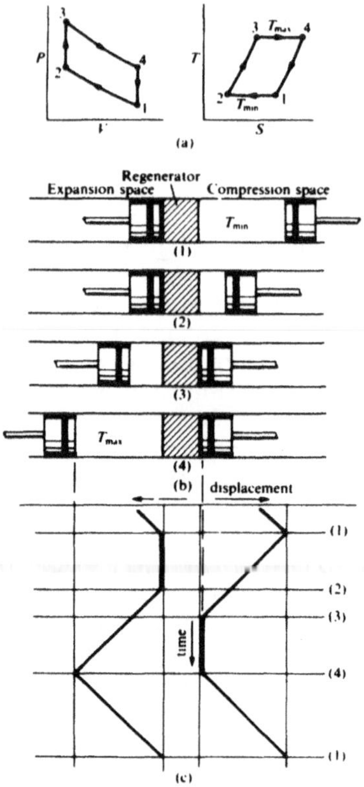

Figure 1.1 The Stirling cycle.

from) the regenerator, so that the volume between them remains constant. Therefore, the working fluid is transferred, through the porous metallic matrix of the regenerator, from the compression space to the expansion space. In passage through the regenerator, the working fluid is heated from T_{min} to T_{max}, by heat transfer from the matrix, and emerges from the regenerator into the expansion space at temperature T_{max}. The gradual increase in temperature in passage through the matrix, at constant volume, causes an increase in pressure.

For expansion (process 3-4), the expansion piston continues to move away from the regenerator towards the outer dead point; the compression piston remains stationary at the inner dead point, adjacent to the regenerator. As the expansion proceeds, the pressure decreases as the volume increases. The temperature remains constant because heat Q_E is added to the system from an external source.

The final process in the cycle is the transfer process 4-1, during which both pistons move simultaneously to transfer the working fluid (at constant volume) back through the regenerative matrix from the expansion space to the compression space. In passage through the matrix heat is transferred from the working fluid, to the matrix, so that the working fluid decreases in temperature and emerges at T_{min} into the compression space. Heat transferred in this process is contained in the matrix, for transfer to the gas in process 2-3 of the subsequent cycle.

The cycle is composed, therefore, of four heat-transfer processes.

Process 1-2 - isothermal compression; heat transfer *from* the working fluid at T_{min} to the external dump.

Process 2-3 - constant volume; heat transfer *to* the working fluid from the regenerative matrix.

Process 3-4 - isothermal expansion; heat transfer *to* the working fluid at T_{max} from an external source.

Process 4-1 - constant volume; heat transfer *from* the working fluid to the regenerative matrix.

If the heat transferred in process 2-3 has the same magnitude as in process 4-1 the only heat transfers between the engine and its surroundings are, (a) heat supply at T_{max} and, (b) heat rejection at T_{min}. This heat supply and heat rejection at constant temperature satisfies the requirement of the Second Law of Thermodynamics for maximum thermal efficiency, so the efficiency, η, of the Stirling cycle is the same as the Carnot cycle, i.e. $(T_{max} - T_{min})/T_{max}$. The principal advantage of the Stirling cycle over the Carnot cycle lies in the replacement of two isentropic processes by two constant volume processes, which greatly increases the area of the P-V diagram. Therefore, to obtain a reasonable amount of work from the Stirling cycle, it is not necessary to resort to very high pressures and swept-volumes, as it is in the Carnot cycle.

The Stirling Cycle Refrigerator

The ideal machine used to describe the operation of a Stirling cycle prime mover, converting heat to work, can also be used as a refrigerating machine. The only difference is that the temperature of the heat supplied during expansion is *lower* than the temperature of heat rejected from the working fluid in compression. This is illustrated in Figure 1.2 where P-V and T-S diagrams for the prime mover and refrigerating machine are superimposed.

When the Stirling-cycle machine operates as a refrigerator, heat is lifted from the cold zone during the expansion process 3'-4'. The work of compression (area 1-2-5-6) is the same for both the prime mover and refrigerator. The work of expansion (area 4'-3'-5-6), in the case of a refrigerator is less than the compression work and the work input is equivalent to area 1-2-3'-4'. During transfer from

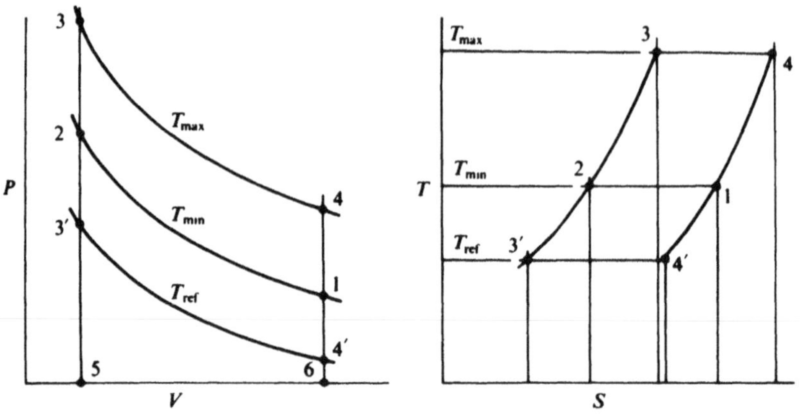

Figure 1.2 The Stirling cycle as prime mover and a refrigerating machine.

the compression space to the expansion space, in process 2-3', the working fluid experiences a *decrease* in temperature, and a corresponding increase in temperature during the alternate transfer process 4'-1.

The performance of a refrigerator is assessed in terms of its coefficient of performance (COP), where

$$\text{COP} = \text{heat lifted/work done} = T_{ref}/(T_{min} - T_{ref}).$$

The Stirling Cycle Heat Pump

The Stirling cycle heat pump operates exactly as the refrigerating machine described above, with the temperature of the expansion space T_{ref} less than the temperature of the compression space T_{min}. The difference between a heat pump and refrigerating machine is that, for the heat pump, both T_{ref} and T_{min} are increased. In the prime mover and refrigerator application, T_{min} is the ambient atmospheric temperature at which cooling water is available. In the case of the heat pump, T_{min} is the temperature, above ambient, at which heat is rejected from the system for heating a concert hall or office building. Heat is supplied to the cycle (at T_{ref}), from atmospheric air, river water or other source at approximately ambient atmospheric temperature.

A comparison of the Stirling cycle as a heat pump and a refrigerator is drawn in Figure 1.3. In both cases, work from an external source is required to drive the cycle, and is equivalent to area 1-2-3-4. In the case of the heat pump, the useful product is the heat rejected at temperature T_{min}, and the performance of a heat pump is therefore assessed as:

$$\text{COP}_{HP} = (\text{heat rejected})/(\text{work done}) = (T_{min})/(T_{min} - T_{ref}).$$

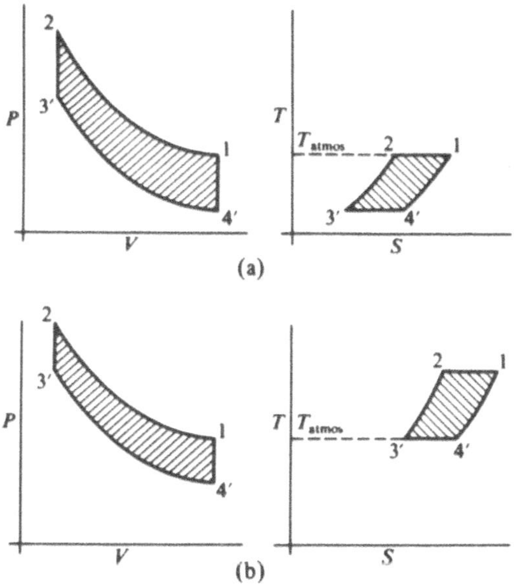

Figure 1.3 Stirling cycle as a heat pump and a refrigerating machine.
a) Heat pump
b) Refrigerating machine

The Stirling Cycle Pressure Generator

For applications where an output of compressed gas, liquid or vapour is required a machine incorporating a regenerator and resembling a Stirling engine can be devised. Working fluid is bled from the cylinder during the high pressure phase and admitted during the low pressure phase. The net result is the conversion of a high temperature heat supply to work in the form of pressurized fluid; it is called a 'thermocompressor' and was invented in the 1930's by Vanevar Bush, a professor at the Massachusetts Institute of Technology.

Valves to regulate the flow of working fluid to and from the cylinder are invariably required in a thermocompressor. Therefore from the definition given earlier the system is an Ericsson rather than a Stirling engine. Nevertheless it is frequently classified as a Stirling engine.

PRACTICAL STIRLING ENGINES

The ideal Stirling cycle discussed above comprised four thermodynamic heat transfer processes, two isothermal and two constant-volume. It was assumed that all processes were thermodynamically reversible, and that the processes of compression and expansion were isothermal, thereby implying infinite rates of heat transfer between the cylinder walls and working fluid. It was also assumed that all the working fluid was in the compression or expansion space during the processes of expansion and compression, so the effect of voids in the regenerative matrix, clearance space, or pockets in the cylinder were neglected. The two pistons were caused to move in some discontinuous fashion to achieve the prescribed working fluid distribution, and all aerodynamic- and mechanical-friction effects were neglected. Regeneration was assumed to be perfect, which implied an infinite rate of heat transfer between the working fluid and regenerative matrix, and an infinite heat capacity of the regenerative matrix.

In practical Stirling engines few of these idealizations can be realised and the thermal efficiency falls well below the Carnot value of the ideal cycle. The actual thermal efficiency may be quoted as a fraction of the theoretical Carnot efficiency. This ratio, called the 'relative efficiency',

$$\eta_{rel} = \text{actual thermal efficiency/Carnot thermal efficiency,}$$

rarely exceeds 0.4 but can reach as much as 0.72 in well designed engines.

To illustrate the ideal cycle, a mechanical arrangement was assumed of two opposed pistons, with intermediate regenerator. The two-piston machine is just one of several different mechanical arrangements to be considered later. A practical version of the two-piston machine is shown in Figure 1.4. It consists of a V engine, with both pistons coupled to a common crankshaft. The spaces above the pistons constitute the compression and expansion volumes and are coupled by a

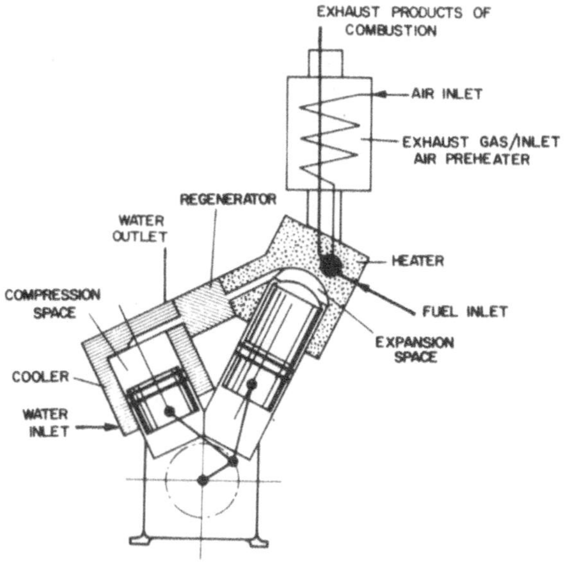

Figure 1.4 Line diagram of vee configuration two-piston Stirling engine.

duct containing the regenerator and additional heat-exchangers.

In the operation of such an engine, a significant departure from ideality arises as a consequence of the continuous, rather than discontinuous, motion of the pistons. This results (as shown in Figure 1.5) in a P-V diagram, which is a smooth continuous envelope.

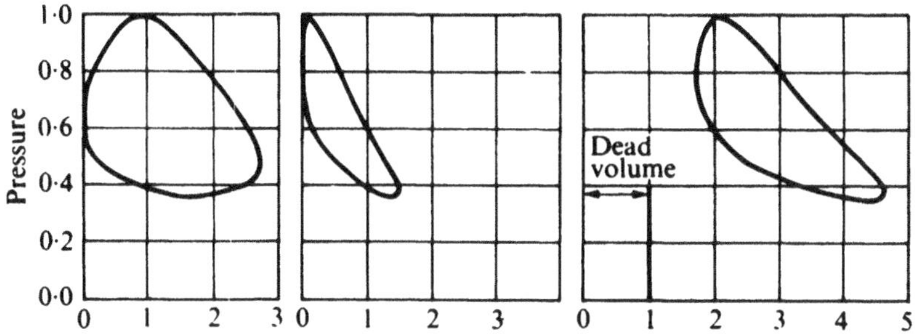

Figure 1.5 Pressure-volume diagrams for the practical Stirling engine.

The four processes of the ideal cycle are not sharply defined. The processes of compression and expansion do not take place wholly in one or other of the two spaces, so that three P-V diagrams may be drawn, one for the compression space, one for the expansion space, and one for the total enclosed volume, which includes the 'dead' space. The 'dead' space is defined as that part of the working space not swept by one of the pistons, and includes cylinder clearance spaces, void volumes of the regenerator and other heat exchangers, and the internal volume of associated ducts and ports. The P-V diagram for the expansion space represents the total positive work of the cycle, whereas the diagram for the compression space represents the compression (or negative) work of the cycle. The difference in the areas of these diagrams is the net cycle output, the 'indicated' work available for overcoming mechanical-friction losses and for providing useful power to the engine crankshaft.

In a cycle where the processes of compression and expansion are isothermal and there are no friction losses, the difference in the area of the expansion- and compression-space diagrams will be found to be exactly equal to the area of the P-V diagram for the total working space. In a practical engine, of course, this equality does not obtain, because aerodynamic-flow losses in the regenerator and other heat exchangers cause differences in the pressure of the working fluid in the compression and expansion spaces. Flow losses are important, because (as shown in Figure 1.6), they cause a decrease in the area of the expansion space P-V diagram, resulting in, (a) a decrease in the net cycle output (and, hence in efficiency) of a prime mover, and (b) a decrease in the cooling capacity and the COP of a refrigerating machine.

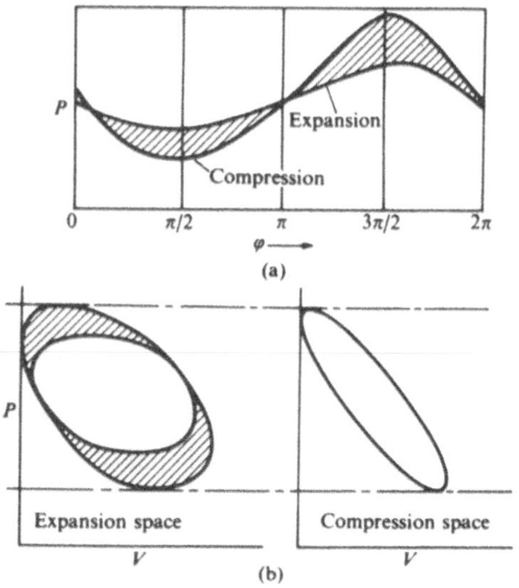

Figure 1.6 Effect of aerodynamic-flow loss on engine work.

The sinusoidal piston motion results in the working fluid being distributed in a cyclically time-variant manner throughout various temperature ranges, and it is not possible to draw a meaningful T-S diagram for the total mass of the working fluid. It is possible to draw T-S diagrams for particular particles of the working fluid, as they move from one temperature range to another, but no convenient way has been found to combine these in multiple diagrams.

Another major departure from ideality is that the processes of compression and expansion are not isothermal. In an engine, running at a reasonable speed (say, 1000 rev/min), it is likely that the processes are more nearly adiabatic (no heat-transfer) than isothermal (infinite heat-transfer). This is a principal factor of the inability of a practical engine to achieve the Carnot efficiency. In order to improve the situation special heat exchangers are often incorporated (as shown in Figure 1.4), including, (a) a heater, adjacent to the expansion space, imparting heat *to* the working fluid, and (b) a cooler, adjacent to the compression space, abstracting heat *from* the working fluid. Despite the advantages of improved heat-transfer, the provision of such heat exchangers imposes some penalties. Additional aerodynamic-flow losses are likely, with the consequent deleterious effect on performance discussed above. The dead volume will be increased by the void volume of the heater and cooler, and this has a critical effect on the performance of regenerative engines. Furthermore, the working fluid is heated, not only when flowing from the regenerator *to* the expansion space, but also when flowing *from* the expansion space to the regenerator. Similarly, the working fluid is cooled when flowing to, as well as from, the compression space. The provision of one-way flow systems is possible, but adds much complication to the machine.

The increased flow-loss and void-volume (along with considerations of cost, size, and weight) combine to produce a comprised heat exchanger design. Consequently, substantial differences may exist between the temperatures at which heating (combustion products) and cooling (water or air) is available and those experienced by the working fluid. This is illustrated diagrammatically in Figure 1.6 and might be considered representative of the temperatures in a fossil-fuelled water-cooled regenerative engine. The temperatures of the combustion products and cooling water are 2800K and 280K, respectively. The metallurgical limit of the materials used for the expansion cylinder and heater is 1000K. This provides a steep temperature gradient, from 2800 to 1000K, between the combustion products and cylinder wall, with the potential for high rates of heat transfer even though the products of combustion may have a low density. Further temperature gradients of (say), 100K, may exist between the mean temperature of the working fluid in the expansion space and the heater. Another, of say 50K, may exist between the working fluid in the compression space and the cooler. The cyclic temperature excursion of the working fluid then varies from (280 + 50) = 330K to (1000 - 100) = 900K, a much more limited range than the superficial difference in the extreme temperatures of 2800 - 280 = 2520K.

Whereas the Carnot (or Stirling) cycle efficiency for the system *might* be calculated as:

$$\eta_C = (2800 - 280)/2800 = 2520/2800 = 90 \text{ percent},$$

to give a more realistic picture it *should* be calculated as

$$\eta_C = (900 - 300)/900 = 570/980 = 63 \text{ percent}.$$

This example demonstrates one of the major difficulties in the commercial application of Stirling engines - one shared by the gas turbine and steam engine but minimized in the internal combustion engine - the question of materials. Some parts of the machine (the heater and expansion space) are exposed, continuously, to a high temperature, and are subject, therefore, to the metallurgical limit of the heater and expansion cylinder materials.

The allowable temperature-excursion of the working fluid in a Stirling engine is limited to a fraction of that permissible in an internal-combustion engine, where the maximum cycle temperatures are attained only momentarily. Thus, although regenerative cycles between given temperature limits are thermodynamically more efficient than Otto- or Diesel-cycles, in practice regenerative engines are compared with gas (or oil) engines operating with radically higher temperature limits. Despite this handicap, thermal efficiencies of the best Stirling engines can be as high as those for a Diesel engine, typically 35 to 40 percent. Future application of ceramic materials will raise the efficiencies of both Stirling and Diesel engines.

Not all the heat available from combustion of the fuel and air can be transferred to the working fluid, and the heat to exhaust in the combustion products of a Stirling engine represents a direct loss. It must be paid for in terms of the oil or gas burned but has served no useful purpose in the engine. Therefore, an important engine accessory is another heat exchanger (the exhaust/inlet air preheater), used to warm the incoming air by heat transferred from the exhaust gas. This can be of the recuperative type or the regenerative type. In the recuperative type, the two fluids, exhaust gas, and incoming air are separated by walls into separate ducts. In the regenerative type, the fluids flow alternately, usually in contraflow, through the same porous matrix. It is important to distinguish between the internal regenerative heat exchanger, incorporated as an integral part of the engine, and the recuperative (or regenerative) heat exchanger, used as an accessory of the engine for inlet air preheating.

The continuous motion of the reciprocating elements, the non-isothermal compression and expansion processes, the limited heat-transfer in cooling and heating devices, the exhaust-stack loss, the increased dead space, and the aerodynamic-flow losses together constitute the principal reasons for the failure of most practical Stirling engines to fulfill their designers' hopes and ambitions. Other causes of disappointment include deficiencies in regenerator operation, high mechanical-friction losses, temperature equalization as a result of relatively massive conduction paths, and fluid leakage owing to imperfectly designed (or imperfectly operating) seals.

The cooling system must handle nearly all the waste heat from the engine for little can be dissipated to exhaust. For engines of given output and comparable efficiency the cooling system of the Stirling engine must handle about twice the heat load of the Diesel engine. This plethora of heat exchangers, heater, cooler, regenerator, exhaust gas/inlet air preheater and large air-cooled radiator is the principal reason, along with relatively complicated control systems, why the most optimistic projections put the capital cost of Stirling engines at least twice the cost of equivalent diesels.

MECHANICAL ARRANGEMENTS

The essential elements of a Stirling engine are two spaces whose volumes are cyclically varied at the same frequency but with a phase difference. Volume variations in the expansion space *lead* those in the compression space. The two spaces are coupled by a duct containing the regenerator and, in most cases, additional heat exchangers to facilitate heat transfer to and from the working fluid.

These simple elements can be combined in a wide variety of ways catalogued by Walker (1980). Most systems use reciprocating elements to achieve the cyclic volume variations but rotary machines are also possible.

Piston-Displacer in the Same Cylinder

Figure 1.7 shows a Stirling engine arrangement having a single cylinder and two reciprocating elements, a piston and a displacer. This is in contrast to the two opposed piston machine used to illustrate the ideal cycle or the two-piston 'Vee' engines illustrated in Figure 1.4.

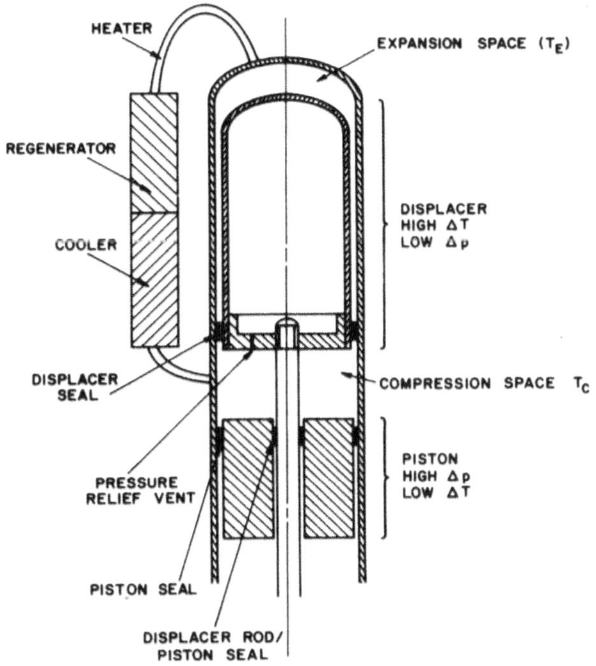

Figure 1.7 Stirling engine arrangement known as the 'piston and displacer in the same cylinder' illustrating the distinction between a piston and displacer.

Pistons

A piston is a heavy structural element with fluid seals to contain the working fluid acting on the topside of the piston, normally substantially higher than the crankcase, buffer space or bounce space pressure acting on the underside of the piston. There is therefore a high pressure difference (Δp) acting on the piston. At the same time the temperature both above and below the piston is generally about ambient so the temperature difference (ΔT) across the piston is zero.

In two piston machines the hot piston is usually equipped

with a dome, called a 'Heylandt crown', to separate the hot expansion space working fluid from the region where the seals operate on water cooled walls.

Displacers

A displacer is a lightweight, essentially no-work, element separating the expansion and compression spaces. Motion of the displacer causes fluid to move from one space to the other. Since the two spaces are coupled through a duct there is no substantial pressure difference (Δp) across the upper and lower faces of the displacer (apart from aerodynamic losses in the heat exchangers and connecting ducts). There is however a substantial temperature difference (ΔT) for the topside of the displacer is exposed to the expansion space temperature and the underside to the compression space temperature. To separate the two spaces, the displacer typically incorporates a Heylandt crown of a length equivalent to two and one half diameters. The walls of the displacer crown and also the walls of the cylinder are made as light as possible to minimize thermal conduction losses along the walls. The displacer is normally filled with some light, low conductivity material or some baffles to minimize internal thermal convection and radiation. In highly pressurized engines it is customary to pressurize the displacer to the mean cycle pressure with working fluid admitted through an extremely fine pressure relief vent. This high-restriction orifice precludes substantial cyclic gas flow to and from the displacer while maintaining the internal pressure close to that of the working space. The strength requirements and the thermal conduction losses of the displacer are therefore minimal.

Displacers are often equipped with seals to prevent passage of the working fluid in the annulus between the displacer and cylinder walls instead of through the path of higher flow resistance in the regenerator and other heat exchangers. In some designs the regenerator is actually incorporated inside the displacer and moves with it.

In other cases no separate regenerator is included in the design and there is no seal on the displacer so the working fluid moves through the annulus of the displacer and cylinder walls in passing from the expansion space and compression spaces. A measure of regenerator action is gained thereby. This concept is called the 'regenerative annulus' and is remarkably effective for small engines (see Walker (1980)).

Advantages of the Piston and Displacer System

A piston and displacer combination instead of two pistons has the advantage of only one pressure seal (on the piston) instead of two (one on each piston). These pressure seals are the Achilles heel of the Stirling engine. A displacer rod seal is necessary but is of small diameter.

Having the piston and displacer operate in the same cylinder is advantageous for the piston and displacer can both 'sweep' the same part of the cylinder (although of course at different times). This is illustrated in Figure 1.8 where section 'E' is the part of the cylinder swept by both reciprocating elements.

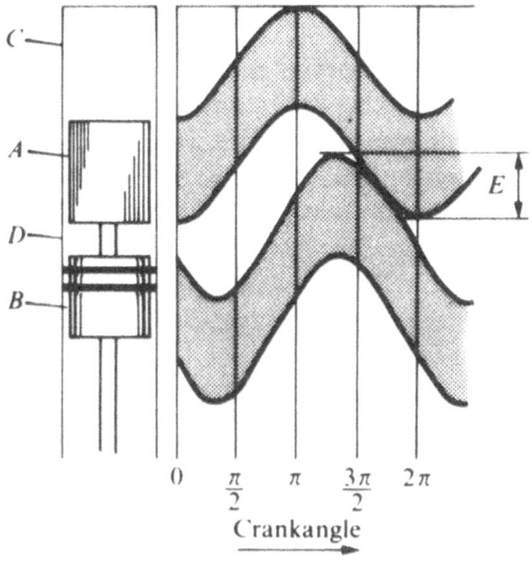

Figure 1.8 Piston and displacer motion in the single cylinder engine.

Piston and Displacer in Separate Cylinders

Another major family of Stirling engines is the arrangement where the piston and displacer are contained in separate cylinders. This is illustrated in the centre diagram of Figure 1.9, showing the three arrangements by which most types of Stirling engines may be classified.

The separate cylinder arrangement has the disadvantage that the compression space is split between the topside of the piston in one cylinder and the underside of the displacer in the other with a connecting duct between. This means the volume of the compression space can never be reduced to zero with consequent reduction in the power output.

One advantage of the arrangement is that the displacer rod seal is moved out of the piston with consequent simplification and reduced seal velocity yielding savings in seal friction and wear.

The separate cylinder arrangement, called the 'split-Stirling'

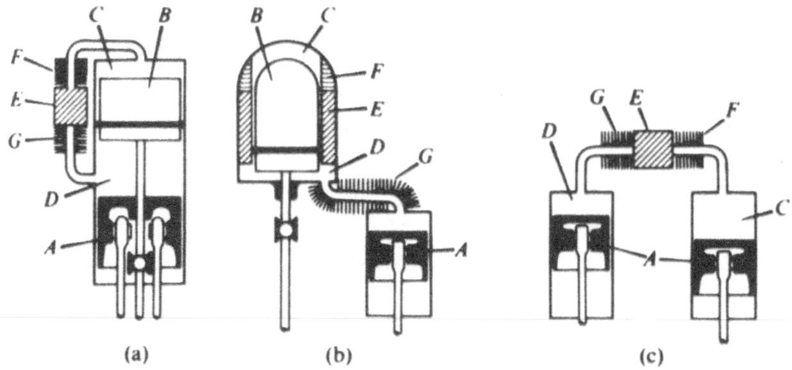

Figure 1.9 Three basic arrangements by which most types of single-acting Stirling engines may be classified.
a) Piston displacer in the same cylinder
b) Piston displacer in separate cylinder
c) Two-piston machine

system is much favoured for the miniature cryocoolers used for infrared night vision rangefinders, gunsights and missile heat seeking guidance systems. The sensitive infrared detectors are mounted on the cold expansion space cylinder head. It is advantageous to locate the small lightweight displacer cylinder and gunsight assembly on gimbals for rapid easy maneuvering and to separate this from the turmoil and hurly-burly of the heavy compressor cylinder complete with crankcase and electric motor drive. This application and free-piston Stirling cryo-coolers are discussed in Chapter 9. See also 'Cryocoolers' (Walker (1983)).

Single-Acting Engines

The engines shown in Figure 1.9 and the innumerable variants thereof are 'single-acting' Stirling engines. They comprise a single Stirling system with two reciprocating elements, a piston and a dis-placer or two pistons operating in one or two cylinders. Engines of a power capacity too large to be accommodated in a single Stirling system can be built as an ensemble of multiple single-acting Stirling systems on a common crankcase. There appears to be no upper limit size. An eight cylinder Vee engine of 800 horsepower is the most powerful engine described in the literature. This was constructed by the General Motors Electromotive Division, LaGrange, Illinois in the late 1960's. However, the largest engine was built much earlier by John Ericsson in 1850. It had four cylinders 14 ft. in diameter and a 5 ft. stroke and operated at 9 revolutions per minute. Journalists of the period describe riding on the pistons of this monster. It was designed as the 450 horsepower engine in a ship called 'The Ericsson' but never achieved this and was subsequently replaced by a steam engine. (John Ericsson was never given to understatement or excessive modesty.) The ship was ill-fated and shortly thereafter capsized in a squall and sank in New York harbour.

Double-Acting Engines

For large engines incorporating several Stirling systems the double-acting arrangement is preferred because the number of moving elements per system can be reduced to one instead of two with single-acting machines.

Two arrangements for double-acting engines are illustrated in Figure 1.10. For systems incorporating only two Stirling systems the two cylinder arrangement devised in 1853 by Franchot may be used. Each cylinder contains a single reciprocating piston. In the cold

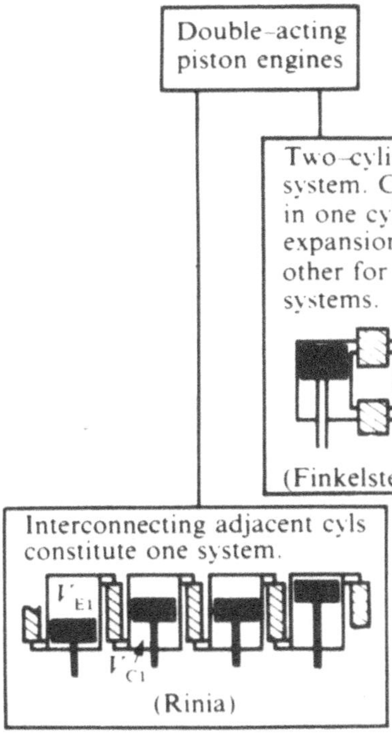

Figure 1.10 Alternative arrangements of double-acting multiple cylinder Stirling engines.

cylinder the spaces above and below the piston are the compression spaces of the two Stirling systems. The hot cylinder contains the two expansion spaces.

If more than two Stirling systems are required the Siemens

arrangement is preferred. The expansion space in one cylinder (above the piston) is coupled through a regenerator to the compression space in the adjacent cylinder (below the piston). Any number of cylinders between three and six may be used; four is a common number and are usually coupled in the 'square four' arrangement shown in Figure 1.11.

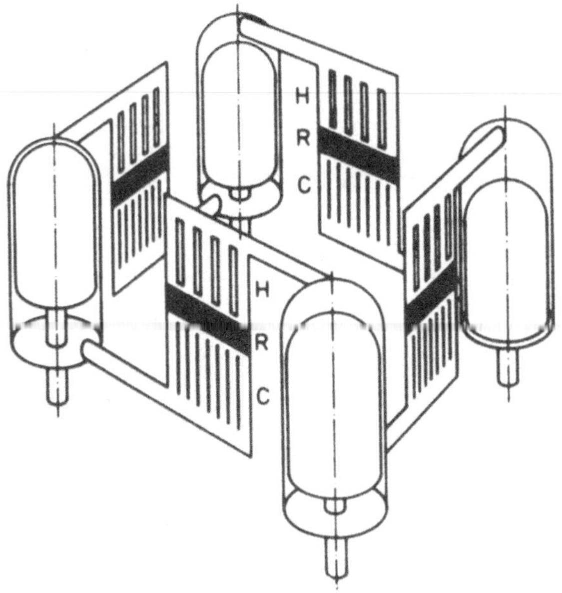

Figure 1.11 Square four arrangement of a Siemens-Stirling engine.

Sir Charles Siemens invented the square four double-acting Stirling engine in 1863, but so far as is known, never reduced it to practice. It was reinvented by Ir. van Weenan in the 1940's in course of the Philips research programme and is now the standard configuration of all large (100 hp plus) Stirling engines.

RECIPROCATOR DRIVES: KINEMATIC OR FREE-PISTON

In all the above variations of Stirling engine systems the reciprocating elements must move in relationship to one another so that volume variations in the expansion space *lead* those in the compression space by approximately 90° or one quarter of the cycle. The actual phase lead is not precise and can vary from 60 to 120° without serious effect on performance.

Kinematic Drive Engines

The reciprocating elements may be connected to a kinematic

drive mechanism such as the crank-slider arrangements illustrated in Figure 1.4. Both the pistons are coupled to a common drive shaft and are caused to move in approximately sinusoidal fashion with a precisely determined phase difference in their motion. Variable phase angle kinematic drive systems can be devised and are one of a number of different control systems that may be used for Stirling engines.

Many different kinematic drive systems are possible, including the simple crank-slider, scotch yoke, wobble-plate, swashplate and rhombic drives. Senft (1982) has given an extensive catalog of the many alternatives available.

Free-Piston Engines

Free-piston engines have no kinematic mechanism coupling the reciprocating elements to each other or to a common rotating shaft. Instead the elements move entirely in response to the gas or other spring forces acting upon them. One well known free-piston Stirling engine arrangement, the Beale engine, is illustrated in Figure 1.12.

Figure 1.12 Beale free-piston Stirling engine.

It consists simply of a heavy piston, and a lightweight displacer operating in a cylinder closed at both ends. The gas in the space below the piston (called the bounce space) acts as a pneumatic spring. It can be replaced by a mechanical spring instead. In either case the spring allows the piston to oscillate on the cylinder at a characteristic frequency called the 'natural frequency' in much the same way that a cork or fishing float 'bobs' up and down in the water.

The motion of the displacer may be sinusoidal as shown by the upper time displacement diagram of Figure 1.12: this is called resonant displacer motion. Alternatively it may follow the more or less square wave motion depicted in the lower time displacement diagram. Here the displacer is said to be 'overdriven', in non-resonant operation and is more graphically described as a 'bang-bang' engine.

We shall look at these and other aspects of free-piston engines in more detail in Chapter 2. Here we simply note the two broad classes of engines: kinematic drive and free-piston.

ADVANTAGES OF FREE-PISTON STIRLING ENGINES

The free-piston Stirling engine is a dynamic, resonant, system operating at a more or less constant frequency and is, therefore, self-starting. When the expansion space is heated up from cold the system requires only a slight, random, perturbation to set it in motion for when heated and not operating the system is in a state of unstable equilibrium, somewhat akin to a ball bearing balanced on the point of a needle. The slightest disturbance will cause the ball to fall or the engine to self-start in its steady vibrating mode. This inherent self-starting capability of the free-piston Stirling engine is very important and a characteristic shared by none of the kinematic drive machines.

The second advantage is that there are no major side thrusts exerted by the reciprocating elements against the cylinder wall. In any simple kinematic drive, i.e. the simple crank-slider mechanism the connecting rod is inclined to the cylinder axis most of the time. The force acting along the connecting rod can be resolved into components acting on the piston along the cylinder axis and also perpendicular to the cylinder axis. This latter 'side-force' pushes the piston against one side of the cylinder and causes uneven wear. Elimination of the piston side-force relieves the lubrication and wear problem sufficiently to permit the use of gas lubricated pistons. The working fluid may therefore be used as the lubricant and so eliminates the problem of regenerator contamination and blockage by the lubricant entering the working space. Elimination of side thrust also facilitates sealing the piston.

Engines exist where the piston is made exceptionally heavy compared to the displacer and the cylinder. In such cases the piston *remains almost stationary* and the displacer and cylinder oscillate. These are called free-cylinder Stirling engines. The cylinder is hermetically sealed during construction thereby eliminating entirely the problems of dynamic seals. This ability to hermetically seal the working fluid within the engine capsule is a major advantage of the free-piston engine concept.

Another important characteristic of free-piston engines is that they are more or less self-regulating. Output work of the engine is extracted from the reciprocating piston (or cylinder in the case mentioned above) to drive a pump, compressor or power generator. The force-stroke diagram of the piston may be somewhat idealized to the rectangular shape ABCD displayed in Figure 1.13 and the area of the figure represents the actual work output of the piston. When the resistance to motion of the driven element is low the piston assumes a

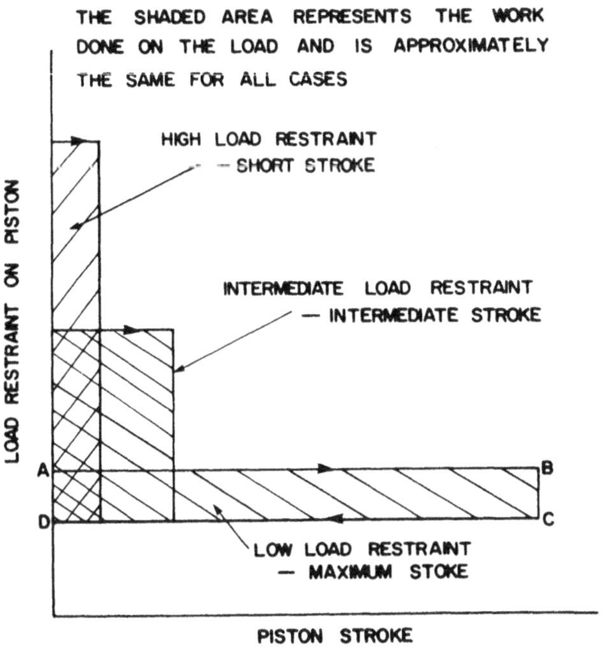

Figure 1.13 Piston force-stroke diagram for a Beale free-piston Stirling engine.

long stroke with relatively little difference between the forces acting on the instroke and outstroke. When the resistance to motion increases the stroke of the piston decreases but the force increases. The changes are such that the *area* of the diagram remains virtually the same. It is virtually impossible to stall a free-piston engine, for a very high resistance to motion imposed on the piston simply results in a very large force exerted by the piston over a short stroke.

DISADVANTAGES OF FREE-PISTON STIRLING ENGINES

The principal disadvantage of the free-piston engine is

simply the lack of a rotating shaft. So many machines in engineering involve rotating shafts that their absence in free-piston engines is sometimes regarded as a disadvantage although in practice the disadvantage is not so great as it appears. Many systems involving pumps, compressors and other machinery include kinematic mechanisms whereby the rotating motion of driven shafts is converted to reciprocating motion. All these can be directly driven by free-piston Stirling engines, and so can a reciprocating pump providing high pressure hydraulic fluid to drive hydraulic motors (e.g. in vehicles).

CLOSURE

In this chapter we have covered a good deal of ground in a superficial way. Readers wishing to secure more detail on any of these matters will find there is a very extensive literature on the subject of Stirling engines. Perhaps the best place to start are the two books we have referred to above, Stirling Engines (Walker (1980)) and Cryocoolers (1983). Both books contain very extensive bibliographies and further discussion of many of the topics introduced above.

REFERENCES

Martini, W. (1980). Private Communication. (see the Stirling Engine Newsletter published by Martini Engineering, Richland, Washington).

Senft, J. (1982). Small Stationary Stirling Engine Design. Proc. Int. Conf. on Stirling Engines: Progress to Reality, pp. 77-84, Inst. of Mechanical Eng., London, (Univ. of Reading, Berks, March).

Walker, G. (1980). Stirling Engines. Oxford University Press, Oxford.

Walker, G. (1983). Cryocoolers. International Monographs in Cryogenics, Plenum Publishing Corp., New York.

CHAPTER 2 - FREE-PISTON STIRLING ENGINES

PART I: ELEMENTARY ASPECTS

HOW A FREE-PISTON STIRLING ENGINE WORKS

Consider a free-piston engine of the type shown in Figure 1.12 and again in Figure 2.1. It consists essentially of three components, a heavy piston, a lightweight displacer and a cylinder sealed at the top end and which may be open or closed at the bottom. A displacer rod of appreciable diameter passes through the piston. The

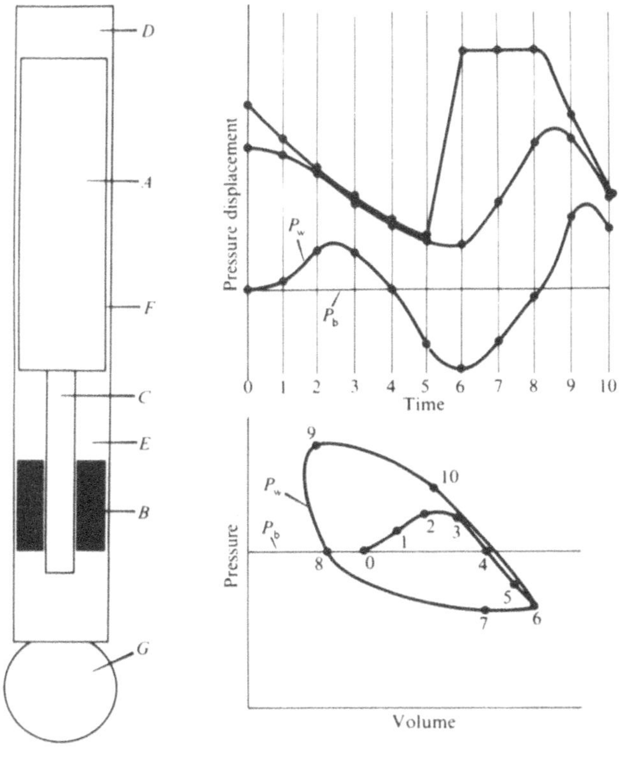

Figure 2.1 Elements on a Beale free-piston Stirling engine. A - displacer, B - piston, C - displacer rod, D - expansion space, E - compression space, F - regenerative annulus, G - boundary space, p_w = pressure in the working space, p_b = pressure in the bounce space (assumed constant), 0 - arbitrary starting condition.

displacer and displacer rod are hollow thin wall shells. They may be open at the lower end so their interior is in fact part of the 'bounce space', the part of the cylinder below the piston. Alternatively, they may be closed and may or may not be provided with a small orifice to allow the interior to be charged to the mean cycle pressure. The 'working space' is above the piston and is divided into, a) the compression space above the piston and below the displacer, and b) the expansion space above the displacer. There is a long, thin annulus between the displacer and cylinder wall and this serves as the regenerative heat exchanger for working fluid passing between the hot expansion space and the cold compression space. Some form of heating is provided for the expansion space and some form of cooling in the compression space.

Let the system be initially at rest in any arbitrary disposition of the piston and displacer as shown in Figure 2.1. The pressure is the same in all the spaces and the temperature throughout is ambient atmospheric. If the bounce space is not closed at the bottom the pressure will of course be atmospheric in all spaces.

Now let the expansion space be heated by the heater. As the temperature in the expansion space increases the pressure, p_w, of fluid in the enclosed working space will increase, process 0-1, on Figure 2.1. As the pressure in the working space increases it acts to move both the piston and the displacer down the cylinder. For this elementary case we assume the bounce space pressure remains constant and further that the piston and displacer rod are equipped with effective seals to prevent leakage of working fluid between the working and bounce spaces.

The force acting on the piston is:

$$F_p = (p_w - p_b)\,(A_c - A_r) \tag{2.1}$$

where
F_p = force acting on the piston,
p_w = working space pressure,
p_b = bounce space pressure,
A_c = cross-section area of cylinder,
A_r = cross-section area of displacer rod.

Recalling Newton's Law of Motion that $F = M\alpha$ where

F = force acting on a body,
M = mass of the body,
α = acceleration of the body,

we may calculate the acceleration of the piston:

$$\alpha_p = \frac{F_b}{M_p} = \frac{(p_w - p_b)(A_c - A_r)}{M_p} \tag{2.2}$$

where α_p = acceleration of the piston,
M_p = mass of the piston.

Similarly the force acting on the displacer:

$$F_d = (p_w - p_b)(A_r) \tag{2.3}$$

and the acceleration of the displacer is:

$$\alpha_d = \frac{(p_w - p_b)(A_r)}{M_d} \tag{2.4}$$

where F_d = force acting on the displacer,
M_d = mass of the displacer.

Therefore the ratio of acceleration:

$$\frac{\alpha_d}{\alpha_p} = \frac{M_p}{M_d}\left(\frac{A_r}{A_c - A_r}\right)$$

Typically (M_p/M_d) is large, say 10:1 and A_c/A_r about 4:1. Substituting these values in the above gives:

$$\frac{\alpha_d}{\alpha_p} = 10 \times \frac{1}{3} = 3.33$$

i.e. the acceleration of the light displacer is over 3 times that of the piston.

Therefore as the pressure in the working space rises above the bounce space due to heating in the expansion space, both elements move along the cylinder but the displacer accelerates more than the piston. This reduces the volume of the compression space between the top of the piston and the underside of the displacer. The working fluid in the compression space is therefore squeezed out of the compression space, through the regenerative annulus into the expansion space. It is heated in the expansion space and so the pressure in the working space increases further between states 1 and 2, consequently increasing the acceleration of the piston and displacer.

Soon, the displacer comes into contact with the piston and moves with it, thereafter. At this stage the compression space has zero volume and no further cold working fluid is transferred from the compression space to be heated in the expansion space and cause further increase in the working space pressure. Nevertheless the working space pressure is, by now, sufficiently above the bounce space pressure to maintain the acceleration and movement of the piston and displacer along the cylinder.

This continued movement of the piston and displacer increases the total working space volume (made up of the void volume of the regenerative annulus and the combined compression and expansion spaces) and with no further transfer of fluid to the hot space the pressure stabilizes (process 2-3) and then begins to fall as the gas expands.

The process of expansion continues until, at point 4, the pressure in the working space equals the bounce space pressure. At this point there are zero gas forces acting on the piston and displacer but the *inertia* of the heavy piston is sufficient to sustain motion along the cylinder causing further expansion of the working fluid and resulting in a decrease of the working space pressure below the bounce space pressure, process 4-5.

With the working space pressure below the bounce space pressure the gas forces acting are reversed and decelerate both the piston and the displacer. The effect is soon manifest on the lightweight displacer which separates from the piston and ascends rapidly to the top of the cylinder (process 5-6). The piston being much heavier is not so easily stopped and continues to move along the cylinder causing further expansion, decreasing the pressure and further increasing the gas forces decelerating the piston. The effect is enhanced by the displacer moving to the top end of the cylinder, reducing the expansion space volume to zero while increasing the compression space to the maximum value. Working fluid is squeezed out of the expansion space, through the regenerative annulus into the compression space. It decreases in temperature during the transfer from the hot to cold space causing a further decrease in pressure in the working space.

The resulting large pressure differential between the working space and bounce space is sufficient to bring the piston to a halt at '6' and then cause it to start ascending the cylinder. The process of compression continues (process 6-7 and 7-8) until the pressure in the working space increases to the bounce space pressure. All this time the displacer is held at the top of the cylinder by gas forces acting on the displacer rod arising from the pressure differential ($p_w - p_b$).

In process 8-9 the inertia force of the piston carries it past the pressure balance point causing a higher pressure in the working space. The displacer starts to fall toward the piston under the action of both gravity and the downward acting gas forces. This induces the working fluid to move from the cold compression space to the hot expansion space, further increasing the pressure in the working space. The downward acting gas forces on the piston act in opposition to the upward inertia forces.

At state 10 the displacer is once more in contact with the displacer, as at point 2, and the cycle is then repeated but without the starting sequence (processes 0-1 and 1-2).

The above is an idealized and simplified explanation of how a free-piston engine works. It is presented here in this way because it is easy to understand and sufficiently realistic for newcomers to the field to be convinced of the feasibility of such a system.

BEALE FREE-PISTON STIRLING DEMONSTRATOR ENGINE

An engine corresponding closely to that described above was developed by William Beale. An early version of the engine operating as an air compressor is shown in Figures 2.2 and 2.3. The compressor

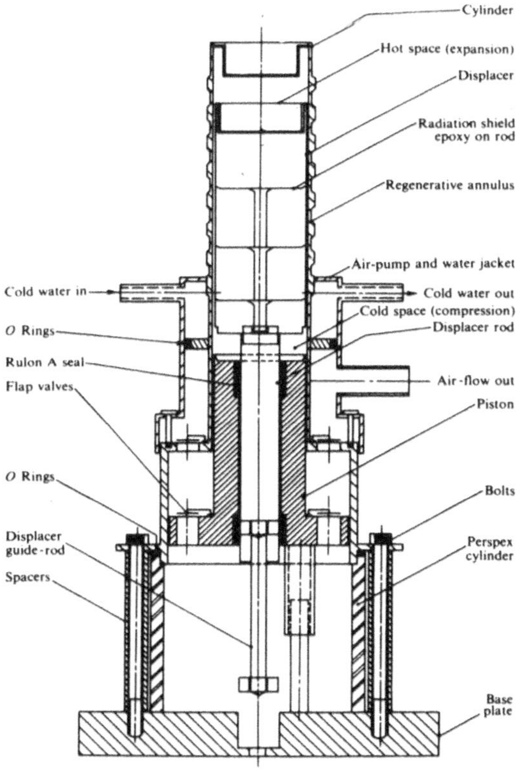

Figure 2.2 Beale free-piston Stirling engine arranged as an air compressor.

cylinder was arranged concentrically about the axis of the working space cylinder with the compressor piston integral with the engine piston.

Agbi (1971) carried out systematic studies of this engine. A characteristic result for the piston-displacer motion, the periodic pressure fluctuation and the overall pressure volume diagram are shown in Figure 2.4.

These results show the piston and displacer moved with approximately sinusoidal motion with the displacer leading the piston by

Figure 2.3 Parts for Beale free-piston Stirling engine arranged as an air compressor.

approximately 50°. The moment of zero expansion space volume (top dead center position of the displacer) was 90° ahead of the zero compression space volume (the point of intercept of the displacer with the piston), a phasing exactly that required for Stirling system operation.

The pressure time curve is misleading as printed for the abcissa is not a true zero pressure. With the abcissa drawn very much lower, at true zero, the pressure ratio would be about 2:1.

This early engine was developed and put into limited production by Sunpower Inc. of Athens, Ohio. Several hundred were manufactured and sold over a decade. Figure 2.5 is a typical production version of the Sunpower demonstration free-piston Stirling engine. It could be had as a linear electric power generator, as an inertia water pump or the air compressor shown earlier. Figures 2.6 and 2.7 are cross-section diagrams of the linear electric power generators and the inertia water pump versions of the engine.

It was supplied with an electric resistance heater clamped around the expansion space but many were adapted to use gas, solid fuel, (wood and coal) or concentrated solar heat.

The engine is no longer in production but the engineering

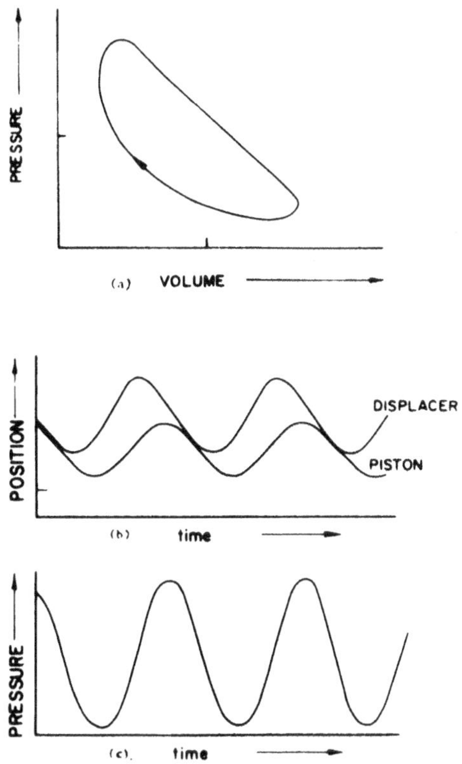

Figure 2.4 Characteristics of the Beale free-piston Stirling engine air compressor, (after Agbi (1971)).

drawings can be obtained* at moderate cost. Newcomers to the field are strongly advised to 'cut their teeth' on free-piston Stirling engines by making and using this relatively simple but well tried version. It is the free-piston Stirling engine that has received the greatest development effort per watt of output and is the one most likely to work for those entering the field. The danger that original thinking will be inhibited by rigid adherence to existing systems should of course be recognized.

TYPES AND CLASSIFICATION OF FREE-PISTON STIRLING ENGINES

There is surprising diversity in the various types of free-piston Stirling engines. Figure 2.8 is a 'family tree' of different arrangements and represents an attempt at orderly classification.

---

*From Sunpower Inc., 6 Byard St., Athens, Ohio, (cost $40 in 1980).

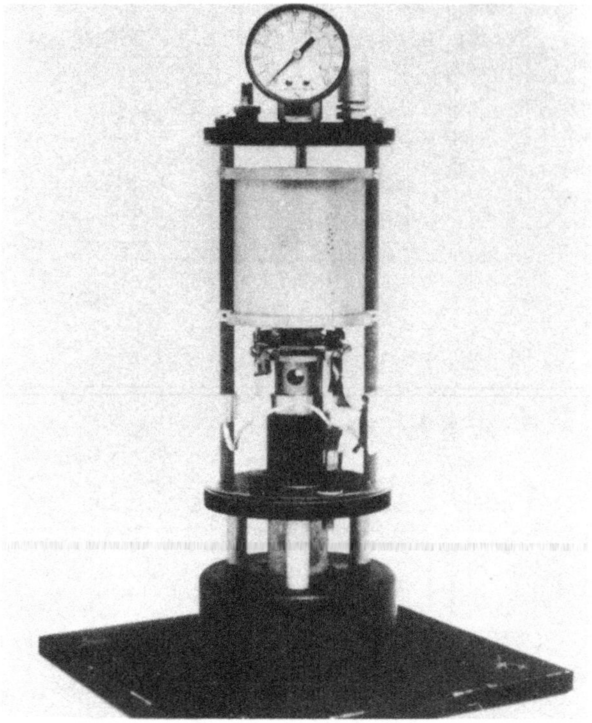

Figure 2.5 Sunpower demonstration free-piston Stirling engine.

The single common feature is that a free-piston Stirling engine has at least one reciprocating element activated only by the fluid forces acting upon it. The motion of this reciprocator can be:

i) simple harmonic in which case the engine is termed a resonant engine, or

ii) non-resonant in which case the engine is described as 'over-driven' or a 'bang-bang' engine.

Single-Acting Free-Piston Stirling Engines

The mainstream of different arrangements shown in Figure 2.8 are engines of the single-acting type. A single-acting Stirling engine has two spaces, a compression space and expansion space, whose volumes are cyclically varied.

Single-acting engines can be classified as:

i) two-piston machines,

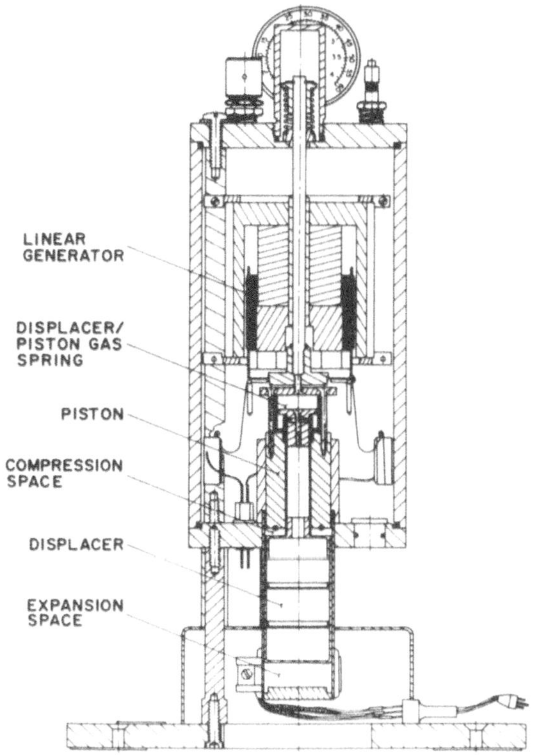

Figure 2.6 Cross-section diagram of Sunpower demonstration engine arranged as a linear electric power generator.

ii) piston-displacer machines; further subdivided into:
   a) tandem cylinder arrangements wherein the piston and displacer are in the same cylinder,
   b) systems wherein the piston and displacer operate in separate cylinders. These are called split-Stirling engines.

i) Two-Piston Arrangements

Two-piston Stirling engines can be devised in any number of different forms. Figure 2.9 shows three different basic forms in which the two pistons oscillate under the combined action of gas pressure variations in the engine spaces V_E, R and V_C and the gas or mechanical springs in the bounce space. An opposed arrangement of twin two-piston engines (Beale (1976)) in perfect dynamic balance eliminates one of the principal problems of free-piston Stirling engines, large pulsations on the engine foundations.

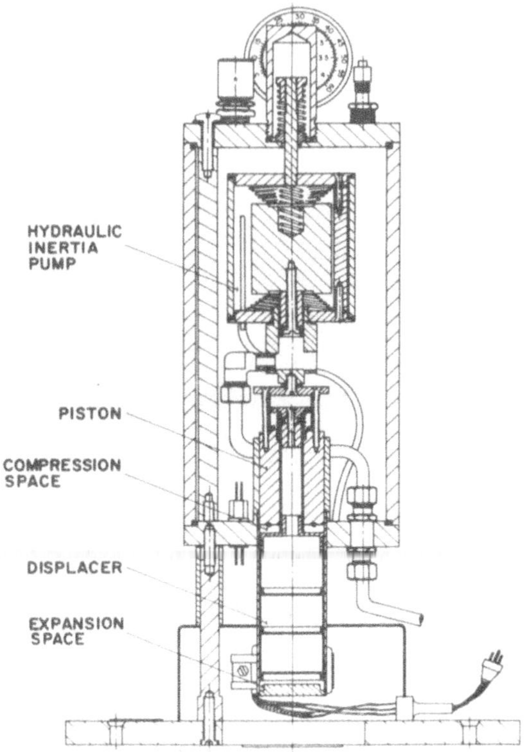

Figure 2.7 Cross-section diagram of Sunpower demonstration engine driving an inertia water pump.

ii) Piston-Displacer Arrangements

Piston-displacer, free-piston, Stirling engines are capable of seemingly infinite variation. Taking first the tandem arrangement of piston and displacer in the same cylinder we have the possibility of a two-mass or three-mass system. All versions of this type of engine actually comprise three separate masses, piston, cylinder and the displacer but in many cases one of the masses is so great as to represent infinite mass compared with the other two. Usually the cylinder is rigidly secured to a solid foundation as the infinite mass so that only the dynamics of the piston and displacer need be considered, i.e. a two-mass system.

There are three groups of two-mass piston-displacer tandem arrangements differing principally in the way in which the displacer is driven. These three groups are illustrated in Figures 2.10, 2.11 and 2.12. When the displacer and piston are coupled by a gas or mechanical spring, as in Figure 2.10, the engine is described as having the 'displacer sprung to the piston'.

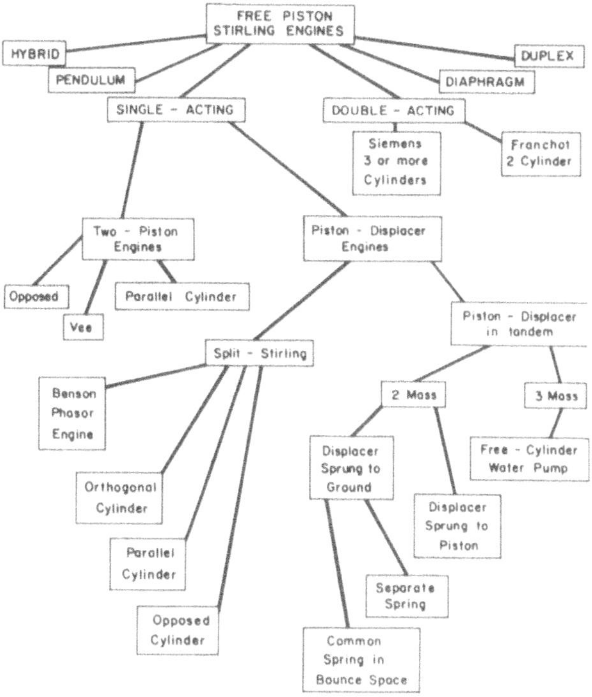

Figure 2.8 Types of free-piston Stirling engines.

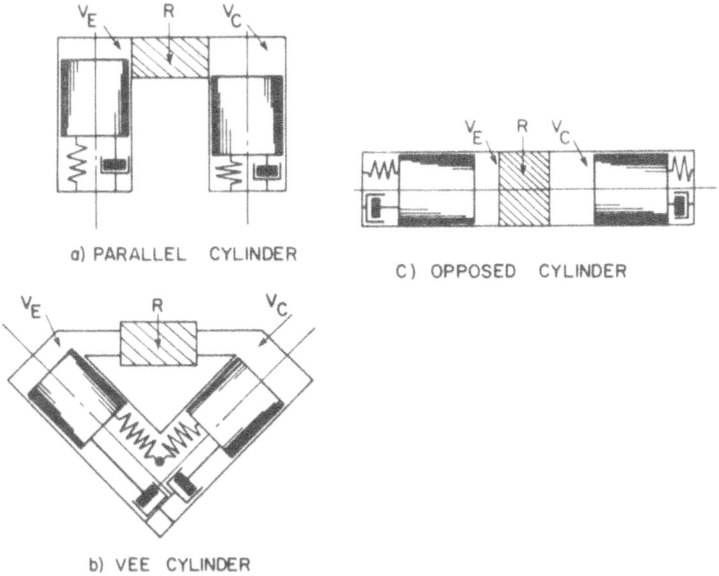

Figure 2.9 Two-piston Stirling engines.

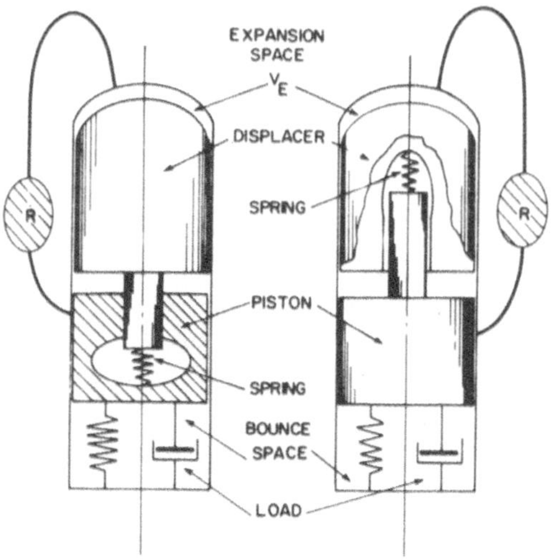

Figure 2.10 Piston-displacer tandem cylinder free-piston Stirling engine with the displacer sprung to the piston.

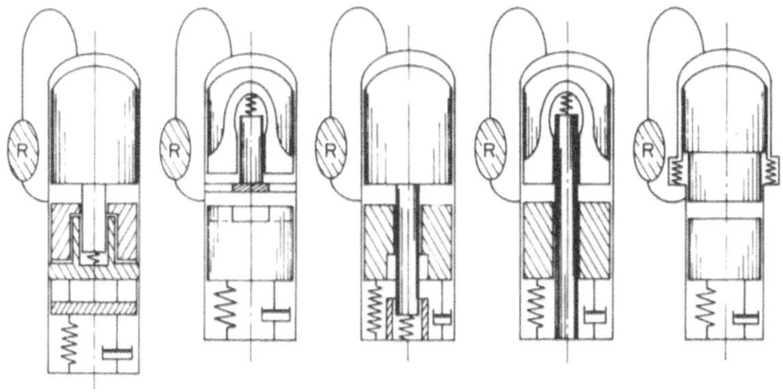

Figure 2.11 Piston-displacer tandem cylinder free-piston Stirling engine with the displacer sprung to ground.

When the 'displacer is sprung to ground' the gas or mechanical spring acts upon the displacer independent of another spring acting upon the piston. Some of the variants of this type of engine are shown in Figure 2.11.

In some cases the displacer is sprung to ground but shares a common gas spring with the piston as shown in Figure 2.12. Because piston and displacer have to operate out of phase this feature of a common spring warrants separate classification.

Three-mass piston-displacer tandem arrangements are most

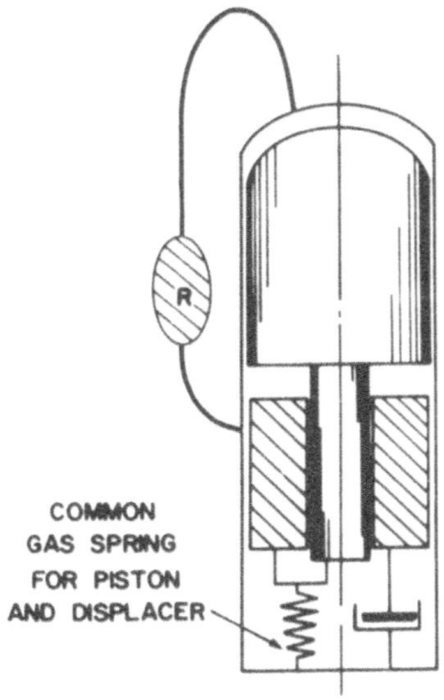

Figure 2.12 Piston-displacer tandem cylinder free-piston Stirling engine with a common spring for the displacer and piston.

commonly found in water pumps, one version of which is shown in Figure 2.13. In these machines the cylinder oscillates to drive the plunger of a reciprocating water pump. The engine is highly pressurized and hermetically sealed during manufacture thereby eliminating all external dynamic seals. The system can be arranged in various ways depending on the displacer drive method. Usually the piston is so massive that its cyclic displacement is small compared with the displacer and cylinder. Water pumps of this type are in production at Sunpower Inc. (Beale (1979)).

iii) Split-Stirling Systems

Split-Stirling engines characterized by the provision of separate cylinders for the piston and displacer exist in many variations. Three possibilities for orthogonal, parallel and in-line axes are shown in Figure 2.14.

The orthogonal cylinder engine has the axes of the displacer and piston cylinders mutually perpendicular and is found in two principal forms. One has the total piston mass divided in two identical half-pistons reciprocating in opposition and each driving a power generator or fluid pump. Figure 2.15 shows a multi-cylinder orthogonal cylinder engine for a hydrostatic drive vehicle, briefly explored by Sunpower in the mid-1970's (Beale (1976)).

The other form of orthogonal engine illustrated in Figure 2.16 is the Benson 'phasor' engine. In this case, work is taken out from one piston only. The other mass, called a 'phasor', and reciprocating in opposition to the piston acts in similar fashion to a

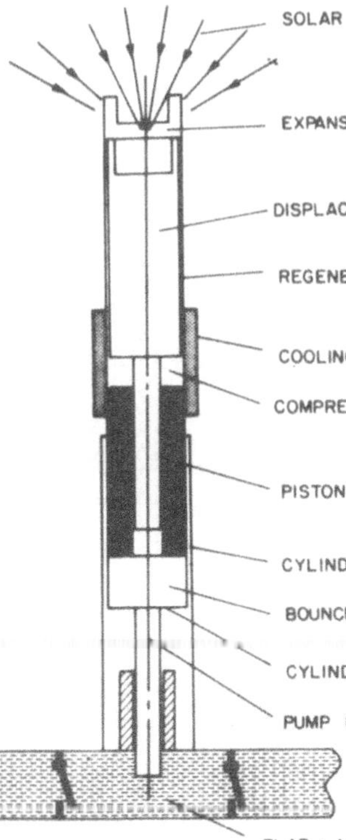

Figure 2.13 Free-cylinder, three-mass, piston-displacer, tandem cylinder, free-piston Stirling engine arranged as a water pump.

flywheel or energy reservoir and allows considerable flexibility in the stroke and loading of the power piston.

Split-Stirling engines are particularly well suited for hybrid operation in which one reciprocating element is driven by a crank and connecting rod and the other by fluid forces only. There are two types of system; one where the piston is the driven element and another where the displacer is the driven element. The case where the piston is the driven element is widely used for Stirling cryocoolers with a miniature displacer cylinder having a tip cooled to cryogenic temperature, incorporated in miniature infra-red night vision and heat seeking missile guidance systems (see Chapter 9).

Another possibility for systems with a driven compressor is the Fluidyne refrigerator. Liquid-piston Stirling engines of the type described by West in Chapter 10 are used essentially for pumping (usually water) utilizing a thermal heat source. However, in principle, they could be used for refrigeration purposes with the oscillating liquid column energized by pressure pulsations produced by a driven compressor. One entertaining concept has the liquid piston of condensed

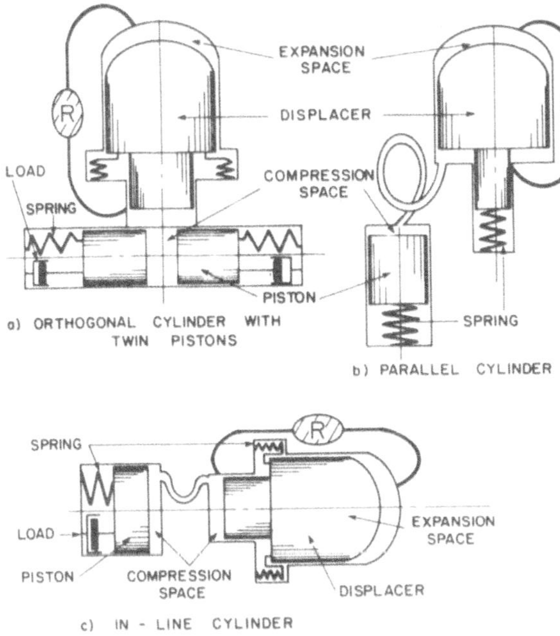

Figure 2.14 Split-Stirling engine.

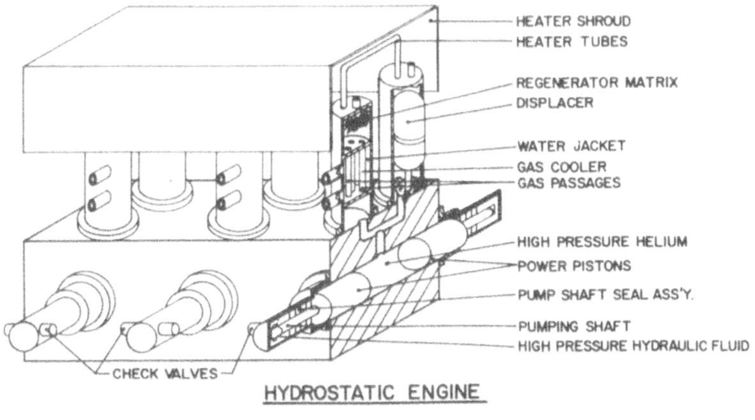

Figure 2.15 Concept for orthogonal cylinder free-piston Stirling engine for hydrostatic vehicle drive system, (after Beale (1976)).

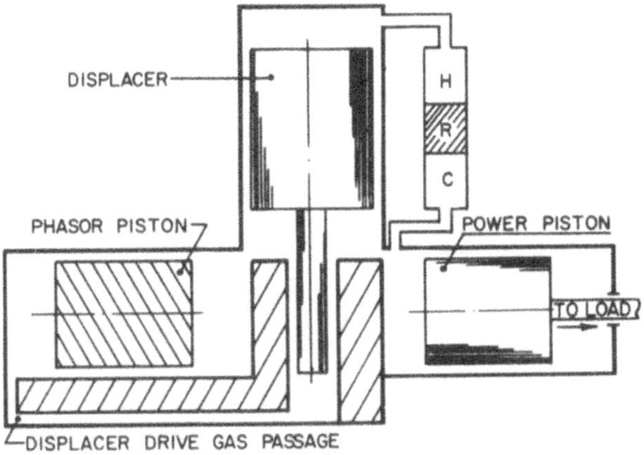

Figure 2.16 Benson phase orthogonal cylinder free-piston Stirling engine.

working fluid resulting from the refrigerator action.

Another hybrid arrangement is feasible where the displacer is driven by a kinematic mechanism and the piston is of the free element. So far as is known this has never been reduced to practice. The work input to drive the displacer would be low for the principal energy input is high temperature thermal energy to the expansion space. The resulting pressure fluctuations, at a frequency corresponding to the displacer drive, act upon the piston causing it to oscillate against the resistance of the gas or mechanical spring. One possible application might be a thermally activated mechanical hammer or compacting device.

Double-Acting Free-Piston Stirling Engines

Double-acting free-piston Stirling engines incorporate multiple engine systems but have only one reciprocating element per system. This is in contrast to single-acting Stirling engines with two reciprocating elements per system.

There are two possibilities for double-acting Stirling engines:

a) engines with two cylinders only, called Franchot engines,

b) engines with more than two cylinders called Siemens engines.

Figure 2.17 illustrates these two concepts. Franchot engines comprise two cylinders with two reciprocating masses M_E and M_C operating about 90° out of phase with the expansion space mass M_E leading the compression space mass M_C. One cylinder contains the two expansion spaces and the other contains the two compression spaces. In the arrangement shown, one piston oscillates as an undamped resonant mass and the other behaves as a highly damped light mass.

When more than two cylinders are involved the Siemens arrangement is used wherein the expansion space of one cylinder (above the piston) is coupled to the compression space of the adjacent cylinder as shown in Figure 2.17(b). A three cylinder arrangement is shown there but four, five and six cylinder versions are possible. With a three cylinder engine the phase angle between piston motion is 120° as shown. With four cylinders it is 360/4 = 90°, with five cylinders 72° and with six cylinders 60°.

Multiple cylinder double-acting free-piston Stirling engines operating as thermally activated refrigerating machines or heat pumps are currently being investigated. In such systems two of, say, three cylinders have the expansion space heated while the expansion space of the third cylinder receives heat at low temperature as shown in Figure 2.18. This concept was first reduced to practice by William Martini (1982).

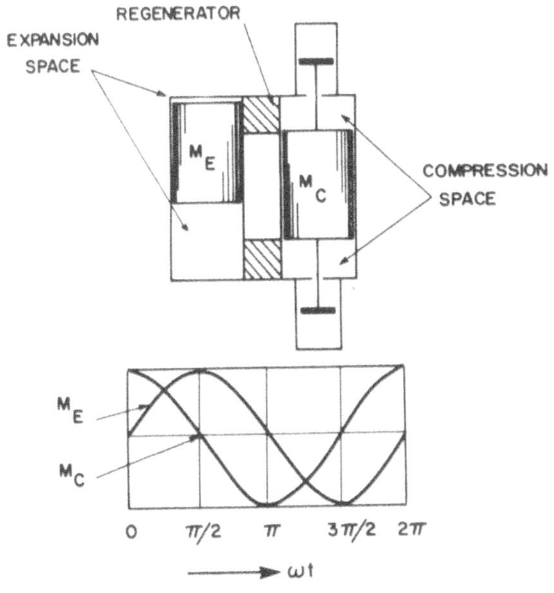

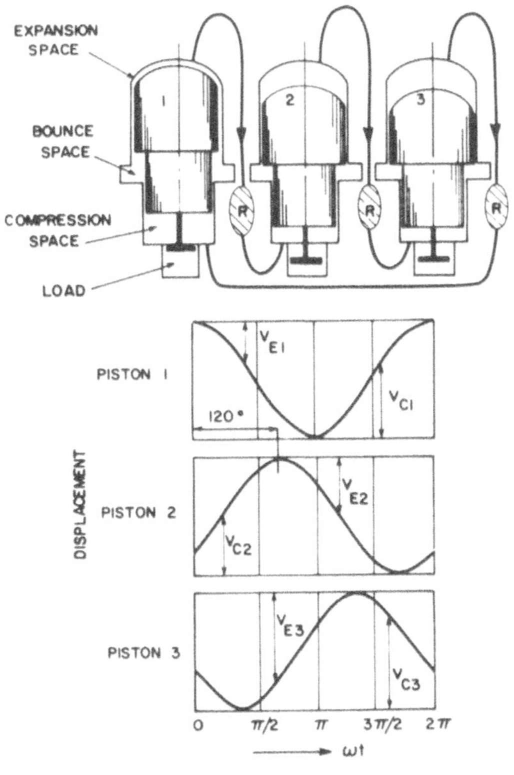

Figure 2.17 Double-acting free-piston Stirling engine.

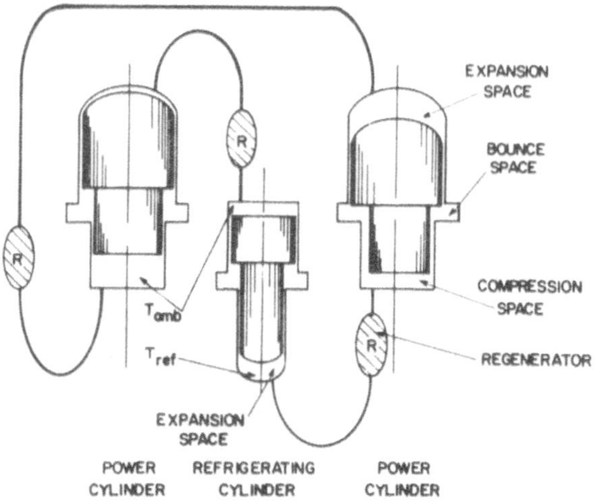

Figure 2.18 Double-acting Siemens Stirling engine with two cylinders as power cylinders and one refrigerator cylinder.

Double-acting Siemens engines can of course be used to power driven refrigerators or heat pumps. In this case the drive system, either a linear electric motor or linear hydraulic drive might be incorporated within the reciprocating element as shown in Figure 2.19. This is called a 'centre-post' arrangement in which the displacer is rigidly constrained except for axial movement by bearings on the

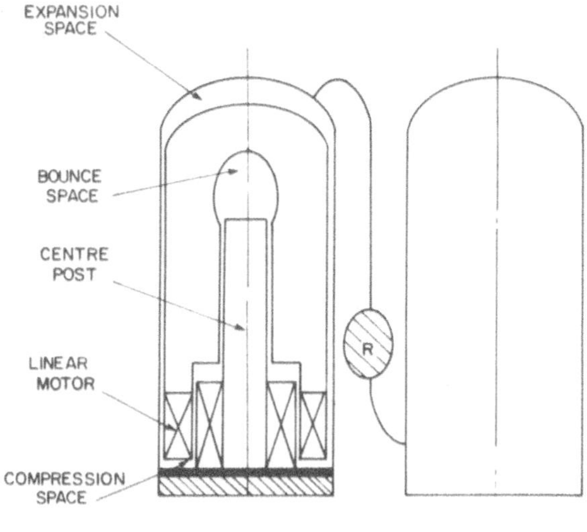

Figure 2.19 Double-acting Siemens Stirling engine refrigerator with linear motor drive of the displacer.

centre-post. These are invariably gas bearings using the engine working fluid as the lubricant.

Hybrid Stirling Engines

Hybrid Stirling engines are Stirling systems in which one of the reciprocators is coupled to a kinematic drive with the other responding to the fluidic forces acting upon it.

We have already considered some aspects of hybrid operation in the discussion of split-Stirling systems but there are many other possibilities. Figure 2.20 shows a tandem arrangement of the crank-

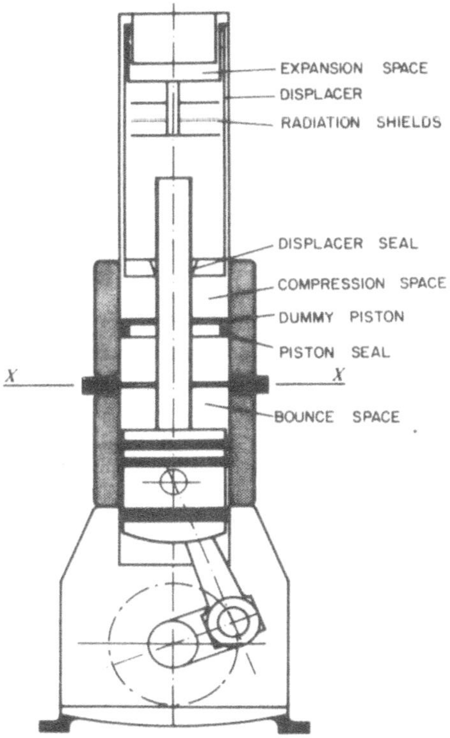

Figure 2.20 Tandem hybrid Stirling engine arrangement.

driven hybrid in which the piston and displacer operate in the same cylinder. In the case shown a dummy piston and displacer rod have been added to the piston of a standard internal combustion engine. An engine of the type, shown in Figure 2.21, was constructed at the University of Calgary in 1970. This arrangement has the great attraction of permitting the use of many existing internal combustion engine parts. One concept for a hybrid tandem arrangement based on

Figure 2.21 Prototype tandem hybrid Stirling engine utilizing industrial engine crankcase and parts.

conventional V8 engine assemblies is shown in Figure 2.22.

Two-piston versions of the hybrid engine, illustrated in Figure 2.23, may be also possible though none have been constructed so far as is known. In the case shown the piston attached to the load is coupled through a crank mechanism to a rotating shaft and the other free-piston simply oscillates in its cylinder displacing gas between the hot expansion space and cold compression space.

Hybrid engines eliminate the greatest disadvantage of completely free-piston Stirling engines, namely the absence of a rotating shaft. They offer the potential of converting existing reciprocating engine and compressor parts using similar technology. Furthermore they permit operation at different speeds instead of the narrow range of operating frequencies, close to resonant, of the free-piston Stirling engine.

The price to be paid for these advantages is high. The introduction of the crank-connecting rod or other simple kinematic drive presages the return of complex piston side forces and the need to lubricate the piston against these. Crankshaft assemblies with dry rubbing bearings tend to have high power consumption, are unable to

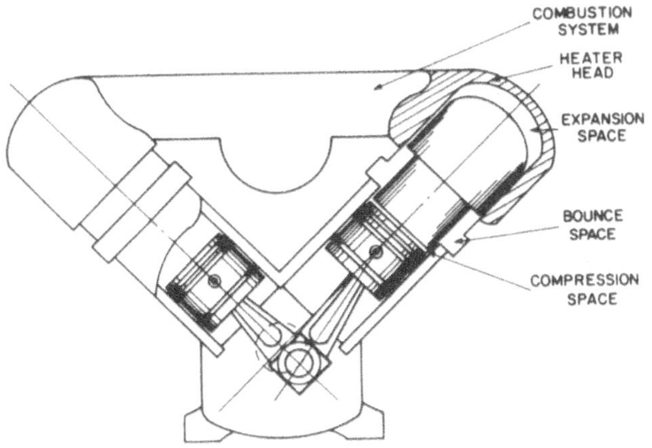

Figure 2.22 Concept for vee hybrid tandem arrangement based on conventional internal combustion engine parts.

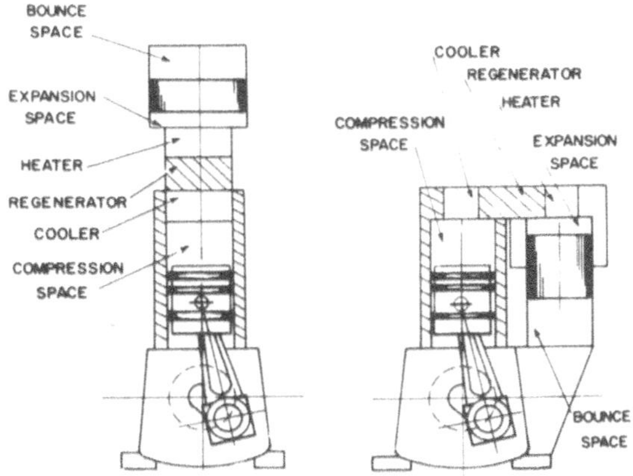

Figure 2.23 Two-piston versions of hybrid free-piston Stirling engine.

support high loadings for long periods and result in considerable detritus. Gas lubricated assemblies are relatively massive. The preferred alternative of liquid lubricant then poses critical sealing problems to prevent the egress of lubricant to the cylinder.

One solution to these problems, under investigation at the University of Calgary, is the use of two-phase two-component working fluids in the engine shown in Figure 2.24. This study is directed to

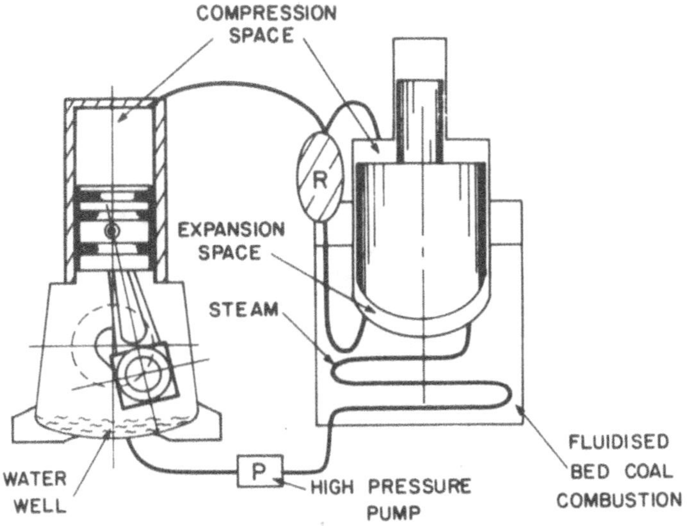

Figure 2.24 Concept for coal-fired two-phase two-component (air-water) working fluid hybrid Stirling engine with water injection and water lubrication.

the use of air and water as the working fluids in very large engines burning coal in fluidized bed combustors. Liquid water is injected at high pressure in the fine bore tube of a single tube boiler immersed in the fluidized bed along with the engine heater tubes. The high pressure steam enters the expansion space and mixes with the gaseous (air) working fluid. Following expansion and passage through the regenerator, where it condenses, the water accumulates in the compression space, riding on the piston and, in passing from the cylinder, serves as the lubricant for the piston rings. The water collects in the sump and is used in the normal way for lubrication purposes for the crankshaft bearings. The principal problem with this scheme is the susceptibility to corrosion of many of the metals normally used in engines. The use of plastic components, plastic coatings and corrosion resistant metals is foreseen.

Hybrid/free-piston Stirling engines have a long history. The very first free-piston Stirling engine was described in a British patent of 1876*. The engine was invented by an Australian mining engineer, Davy Postle of Stanthorpe, Queensland and is illustrated in Figure 2.25. It was intended for use as a refrigerator for marine use on meat ships carrying chilled mutton and beef to England from Australia through the newly constructed Suez Canal. The operating cycle of the engine is described in detail by Walker (1983).

---

*British Patent No. 709, 1873 (26th Feb.), Producing Cold for Preserving Animal Food, W.H. Thomas, for Davy Postle.

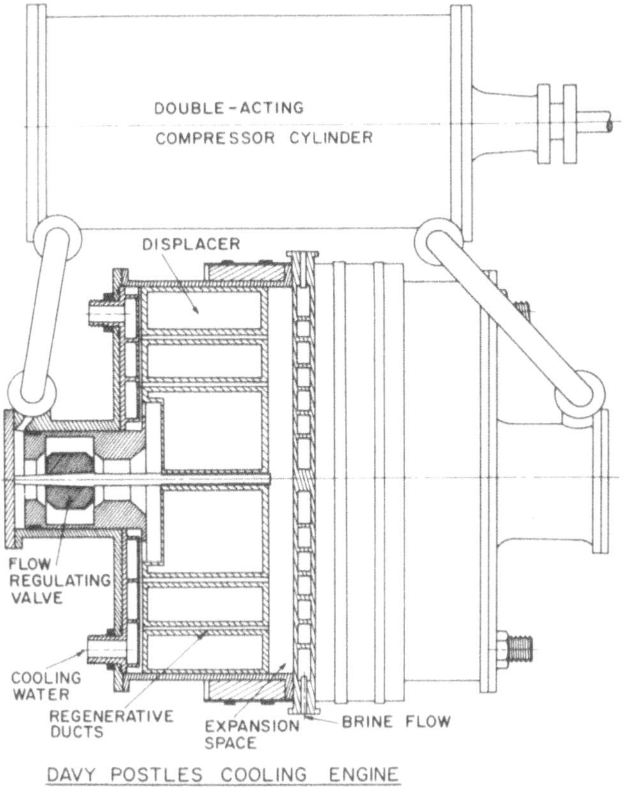

Figure 2.25 Davy Postle's free-piston engine of 1876. Perhaps the original free-piston Stirling engine.

Another early hybrid free-piston Stirling engine was described by Ossian Ringbom in 1905*. Ringbom was a Russian, living in Finland, at the time he took out the U.S. patent. No further details of his work are known, but the engines of the Ringbom type are described by Senft in Chapter 5 herein.+

Pendulum Free-Piston Stirling Engines

Pendulum engines have a long and honourable history principally in connection with clockwork mechanisms activated by a falling weight. There has been very little work done on thermally activated pendulum engines and nothing so far as is known on pendulum Stirling engines until their conception by Horace Rainbow in 1978. Rainbow is

*Ringbom, Ossian, U.S. Patent No.

+I would very much appreciate any information readers may be able to give me about Ossian Ringbom or Davy Postle and their works. G.W.

an engine designer *par excellence* of Bristol, England. One version of the Rainbow pendulum Stirling engine is shown in Figure 2.26. It consists essentially of a simple pendulum composed of a heavy mass swinging at the lower end of the pendulum arm. A piston is attached to the

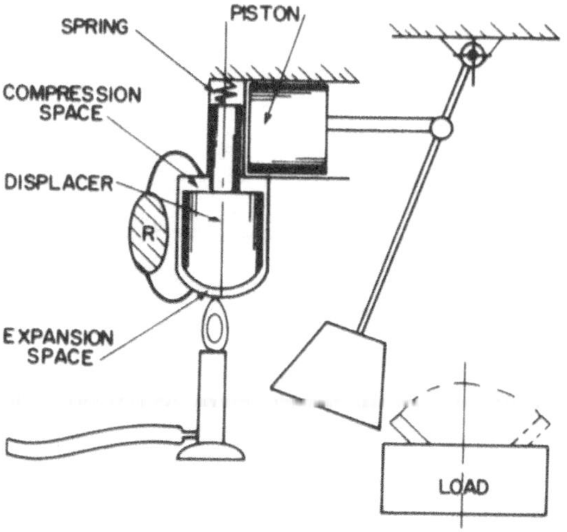

Figure 2.26 Concept for Rainbow pendulum free-piston Stirling engine.

pendulum arm and is thereby caused to oscillate in the cylinder in sympathy with the pendulum motion. The cylinder is rigidly attached to the support and coupled to the vertical displacer cylinder, heated at the lower end. The displacer is a free displacer 'sprung to ground' and oscillates in the cylinder at the pendulum frequency. Events occur so that the displacer is at the top of the cylinder with the air in the hot space and therefore at high pressure when the pendulum swings from left to right, the expansion stroke. During compression stroke when the pendulum swings from right to left the displacer descends, the fluid moves into the cold space above the displacer and the pressure is low. Work can be extracted from this heat engine by means of a fluid pump coupled to the pendulum arm or by a reciprocating device actuated by periodic impact of the pendulum.

In the form shown the engine is a split-Stirling piston-displacer engine in which the pendulum replaces the mechanical or gaseous spring of the piston cylinder or the piston kinematic drive of the hybrid engine. There are of course many other possible variations on themes we have discussed earlier.

Pendulum engines are attractive as low-technology water pumping devices for developing countries. The only machined items necessary are the piston and displacer and cylinder assemblies. Figure 2.27 is an impression of a Rainbow pendulum engine used in this way. The pendulum and A-frame are simple structures made from locally

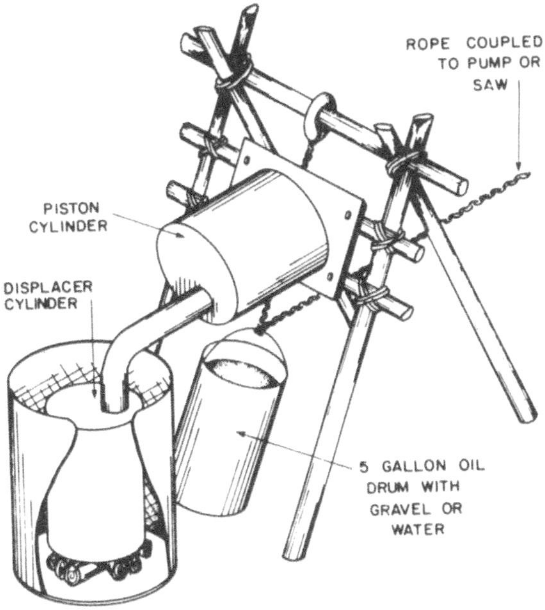

Figure 2.27 Low technology application of Rainbow pendulum free-piston Stirling engine.

available materials and incorporate the machined piston-displacer cylinder assembly.

Diaphragm Free-Piston Stirling Engines

Extreme difficulty sealing the working space against the egress of working fluid and the ingress of lubricant or other contaminants have led designers to the use of flexing diaphragms and metal bellows.

These components do not eliminate all difficulties but merely exchange one problem for another, namely how to fabricate a long-lived flexure. Over the past decade intensive effort has resulted in a new understanding permitting construction of diaphragm engines with very long lifetimes. This new technology of long-lived diaphragms, is briefly discussed in Chapter 6 by E.H. Cooke-Yarborough and Chapter 7 by W. Martini.

The diaphragm Stirling engine developed by Mr. Cooke-Yarborough is illustrated in Figure 2.28. A full description of this engine and its evolution is given in Chapter 6.

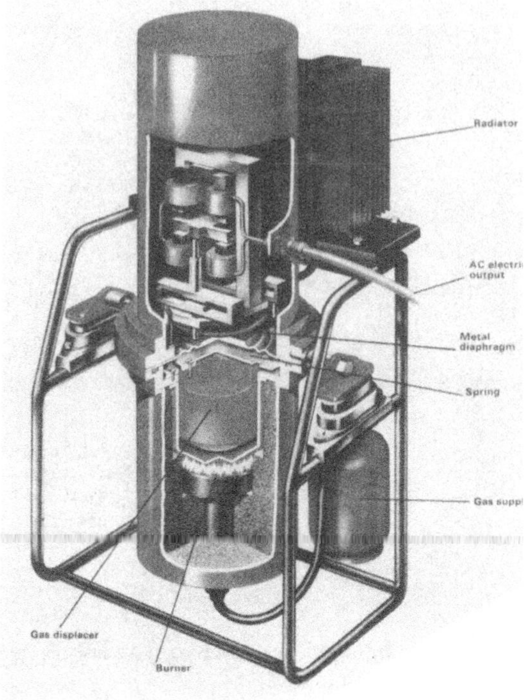

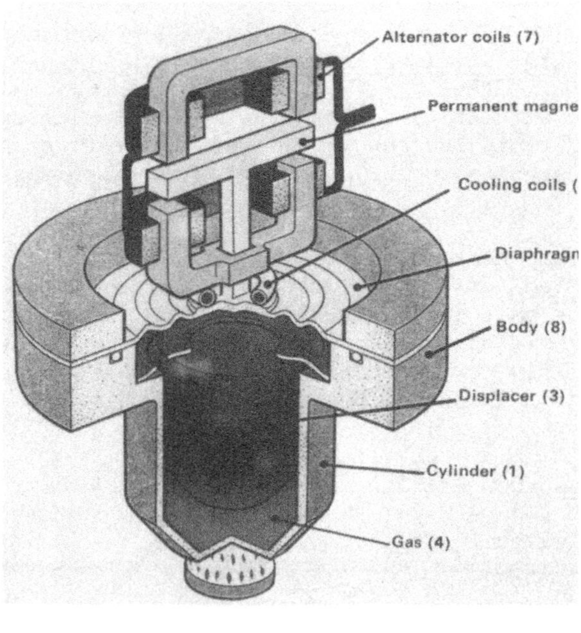

Figure 2.28 The Harwell thermomechanical generator.

Duplex Free-Piston Stirling Engines

We have seen how the Stirling engine can be used to produce power from a given high temperature heat source or to produce refrigeration with an input of work. It requires little imagination to conceive a duplex arrangement wherein an input of heat at high temperature is converted to work in a Stirling engine prime mover and the work then used to drive a Stirling engine refrigerator. A line diagram of the system showing the temperature regimen and energy flows is given in Figure 2.29.

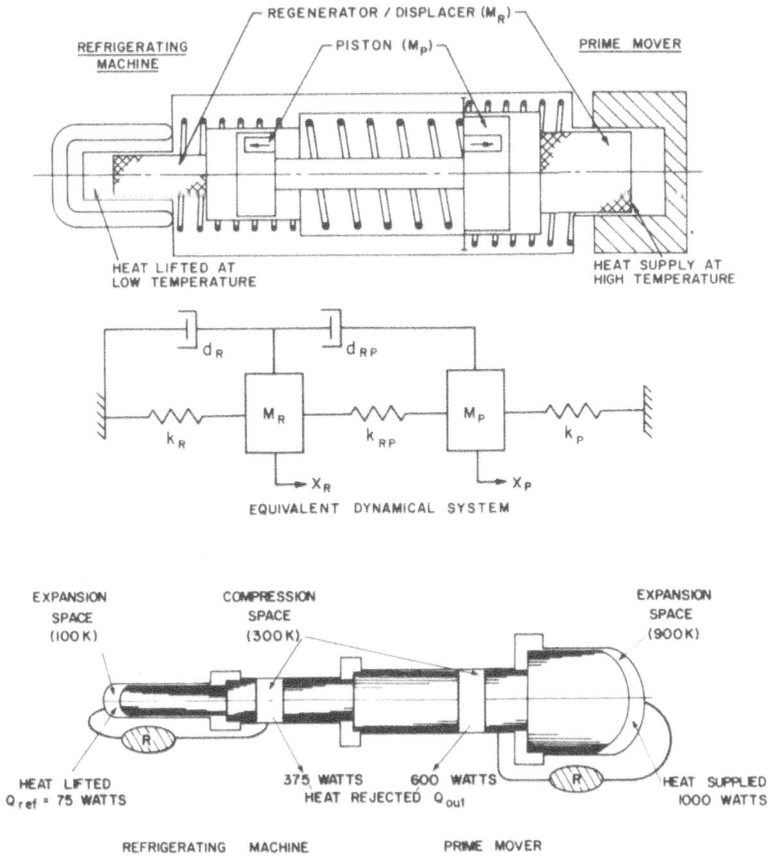

Figure 2.29 The duplex Stirling free-piston engine.

In understanding the duplex Stirling engine it is necessary to appreciate the distinction between it and the Vuilleumier cycle heat activated refrigerator. In the Vuilleumier cycle a large displacer shuttles back and forth in the hot cylinder displacing gas between the hot space and the cold space thereby creating a cyclic change in pressure. A second smaller displacer operating in a separate cylinder also shuttles back and forth at the same frequency but about 90° out of phase with the hot displacer. The result is that a refrigeration effect is created in the smaller cold cylinder. Vuilleumier refrigerators have been intensively developed as cryocoolers over the past 20 years, particularly by the U.S. Air Force. The system is described in detail by Walker (1983).

Vuilleumier machines have a much reduced range of pressure variation compared with Stirling engines and consequently are larger, heavier and less efficient than Stirling systems.

The duplex Stirling engine cryocooler operating as a free-piston engine was conceived by Walker* in the early 1960's but was not reduced to practice until very much later, in 1978, by William Beale and the Sunpower staff. A photograph of the first successful free-piston duplex Stirling engine is shown in Figure 2.30.

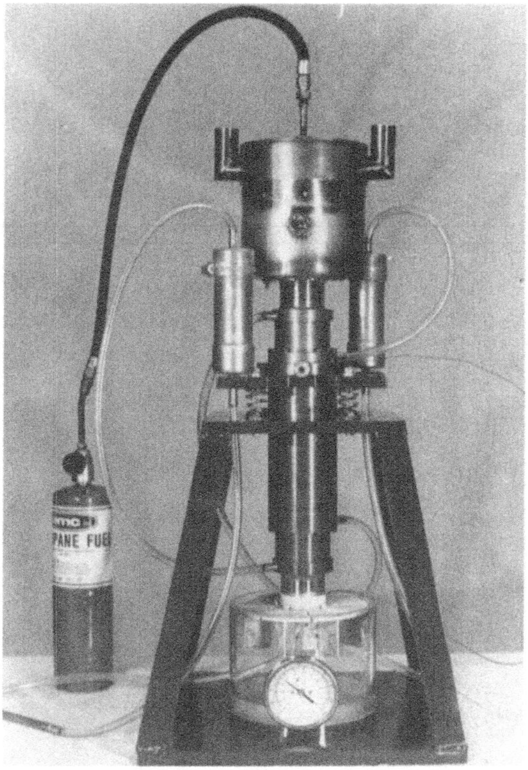

Figure 2.30 The original duplex Stirling free-piston engine, (courtesy Sunpower Inc.).

*Walker, G., (1963), IITRI Chicago Invention Disclosure.

PART II - DYNAMICS OF FREE-PISTON STIRLING ENGINES

INTRODUCTION

For many years the recognized specialist fields of mechanical engineering have included solid mechanics on one hand and thermodynamics, fluid mechanics and heat transfer on the other. It is generally true that specialists in one field rarely feel comfortable working in the other. This is an appreciable impediment to the development of free-piston engines for the essence of their operation is the intimate combination of solid mechanics and thermofluid/heat transfer technology. The difference in temperature of the expansion and compression spaces and the motion of the pistons and displacers together determine the variation in pressure of the working fluid from thermofluid considerations. Yet it is this very change in pressure which disturbs the static equilibrium of the system moving the pistons and thereby causing further change in the pressure. Resolution of this complicated interdisciplinary action is difficult and best approached numerically using the digital computer. We shall leave this to the next Chapter and here will simply idealize our system to understandable concepts with which to form mental images about the operation of free-piston Stirling engines.

Very little is included in this book about the thermofluid/heat transfer aspects of Stirling engines. Such material is discussed at length in the companion volume (Walker (1980)). Those from the solid mechanics field will find it more useful to refer there rather than continue here for they will find this section trivial. It is included for those, like the author, coming from the thermofluid field who find vibration theory a difficult subject to handle. Moreover, it is intended simply as a refresher for those who already have the benefit of an engineering undergraduate course in dynamics. Readers without this foundation should consult the many excellent specialist texts (den Hartog (1956), Tse *et al* (1978), Myklestad (1956)).

VIBRATING SYSTEMS: AN ELEMENTARY REVIEW

Degrees of Freedom

Consider the damped mass-spring system shown in Figure 2.31. If mass 'M' is constrained to move only in the vertical direction a single coordinate, x(t), is required to completely define the location of the mass from the static equilibrium position. If only one coordinate is necessary the system is said to have *one degree of freedom*. Systems are described as having the number of degrees of freedom corresponding to the number of coordinates required to specify the location of the masses involved.

In the equilibrium state the mass 'M' will hang motionless

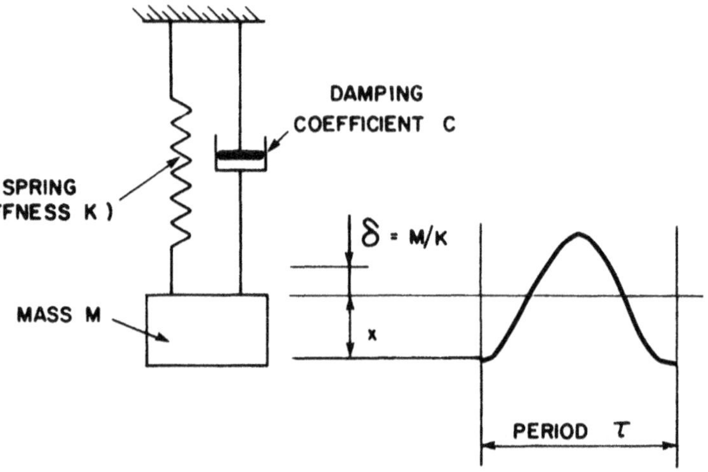

Figure 2.31 Simple damped spring mass system with one degree of freedom.

supported by the spring. The spring will be stretched by an amount δ necessary to generate a force acting upward on the mass equal in magnitude to the gravitational force acting on the mass, the weight. We define the spring stiffness or spring constant 'K' as the force necessary to stretch the spring by one unit of length. Therefore $\delta = Mg/K$, where g is the gravitational acceleration.

The damping device or dashpot exerts no force on the mass when the system is in static equilibrium.

Free Vibration

Now let the mass be pulled down a distance x and suddenly released. At the moment of release the upward spring force on the mass is Kx and acts to cause the mass to move upward toward the original equilibrium position. When the mass reaches the original equilibrium position the spring force will be reduced to zero but, due to inertia, the mass will continue to ascend thereby compressing the spring and generating a negative spring force acting downward on the mass and complementary to the gravitational force. Upon compressing the spring by an amount equal to or less than x the mass would halt and then start to fall back to the original static position and beyond extending the spring towards the point where it was earlier released. This cycle will continue until motion eventually dies away because of frictional damping.

With cyclic action the *period* τ is the time for the periodic motion to repeat itself. The *frequency* f is the number of cycles per unit time. *Free vibration* is the periodic motion that ensues when the system is displaced from its static equilibrium position.

Oscillations occur at a characteristic frequency called the *natural frequency*.

Simple Harmonic Motion

The simplest form of periodic motion of the mass is called *simple harmonic motion* (SHM). This occurs in the spring-mass system when there is no damping. The acceleration of the mass is always proportional to the distance of the mass from a fixed point (the static equilibrium position) on the path and is always directed to that point.

The *instantaneous displacement* x of the mass from the fixed point may be represented by the equation:

$$x = X \cos \omega t$$

where X is the maximum displacement of the mass, ω is a constant and ωt an angle measured in radians. The angular period of the above function is 2π so that $\omega\tau = 2\pi$ where τ is the period in units of time. Therefore $\omega = 2\pi/\tau = 2\pi f$ where f is the cycle frequency. Hence, $f = 1/\tau = \omega/2\pi$ and $\tau = 1/f = 2\pi/\omega$.

If the displacement of the mass M is given by:

$$x = X \cos \omega t$$

then the *velocity* of the mass will be given:

$$v = \frac{dx}{dt} = \dot{x} = - X\omega \sin \omega t = X\omega \cos (\omega t + \frac{\pi}{2})$$

This is a harmonic function with the *same frequency* as the displacement and an amplitude ω *times as large*. The velocity has a phase angle 90° or $\pi/2$ *ahead of the displacement.*

The acceleration of the mass M is given by:

$$a = \frac{dV}{dt} = \frac{d^2x}{dt^2} = \ddot{x} = - X\omega^2 \cos \omega t = X\omega^2 \cos (\omega t + \pi)$$

The amplitude of the acceleration is ω^2 *times the displacement* and its phase angle is 180° *or* π *ahead of the displacement.* These relationships of displacement, velocity and acceleration are displayed in Figure 2.32.

Viscous Damping

All vibrating systems will experience some friction or resistance termed *damping* acting to slow down the motion and cause the

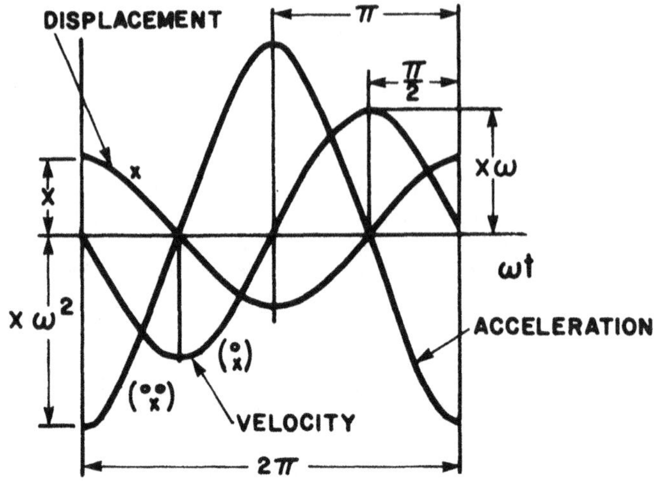

Figure 2.32 Cyclic representation and phase relationship of displacement, velocity and acceleration of mass M in simple spring-mass system.

oscillation to die away. Damping arises from a variety of causes, air damping, fluid friction, Coulomb dry friction, magnetic damping or internal hysteresis.

With heavy damping the motion is not oscillatory. When disturbed the mass simply returns slowly to the static equilibrium position and the system is said to be *over-damped*. A *critically damped* system is the intermediate case. On release the mass does not oscillate and simply returns quickly to the static equilibrium position.

Damping forces always resist the motion of the mass and for simplicity viscous damping is frequently assumed where the resisting forces are proportional to the velocity. This arises when the resisting force is due to the viscous resistance in a fluid medium as in an ideal dashpot. The dashpot is characterized by a proportionality factor 'C', called the *coefficient of viscous damping* and the total damping force at any time is given by the expression $C\dot{x}$ where $\dot{x}$ is the velocity of the mass.

Equation of Motion

The motion of the mass is controlled by Newton's Second Law of motion $\Sigma F = Ma$ where ΣF is the summation of all forces acting on the mass and a is the acceleration of the mass, $\ddot{x}$. Application of Newton's law is manifest in the equation of motion for the mass:

$$M\ddot{x} + C\dot{x} + Kx = 0$$

For free vibration of an under-damped system the motion of the mass as a function of time will be similar to that illustrated in Figure 2.33. It is the motion that results from an impulse or sharp blow applied to a system initially at rest and is given by the solution to the above equation.

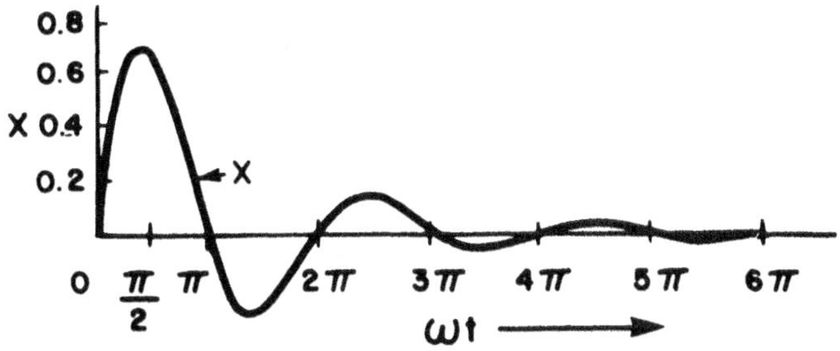

Figure 2.33 Displacement time diagram for damped spring-mass system.

When external forces, usually periodic of the form $F(t) = F_o \sin \omega t$ or $F_o \cos \omega t$, act upon a system the resulting motion is described as *forced vibration.* Initially the system tends to vibrate at its own natural frequency as well as respond at the frequency of the excitation force. However due to frictional damping the natural frequency component decays and only the motion resulting from the excitation force will remain. The system is then in *steady state vibration.*

When the frequency of excitation corresponds to the damped natural frequency the system is in *resonance* and the amplitude of vibration increases to very high values governed only by the damping present in the system. This is illustrated in Figure 2.34 showing the displacement of the mass at different frequencies for different values of viscous damping.

The displacement is given as the ratio $\mu = Xk/F_o$, called the *amplification factor* or *magnification factor.* This is the ratio of actual displacement X, expressed in terms of the deflection of the spring induced by static application of the force F_o, i.e. static deflection = F_o/K.

The horizontal ordinate of Figure 2.34 is the *frequency ratio* r defined as the:

$$r = \frac{\omega}{\omega_n} = \frac{f}{f_n} = \frac{\text{forced frequency}}{\text{natural frequency}}$$

The viscous damping factor ρ is defined as:

$$\rho = \frac{C}{C_{cr}}$$

where C_{cr} is the value of the coefficient of viscous damping when the system is critically damped. C_{cr} can be shown to be: $C_{cr} = 2M\omega_n = 2K/\omega_n$ since $\omega_n = 2\pi f_n = \sqrt{K/M}$, the undamped natural angular frequency.

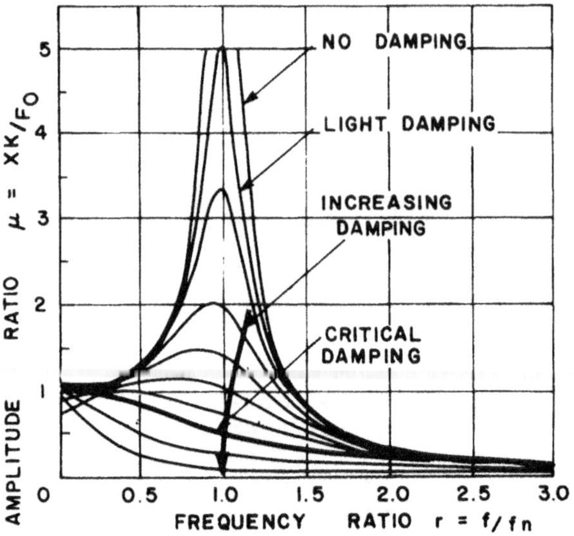

Figure 2.34 Amplitude of oscillation of the mass in a spring-mass system as a fraction of excitation frequency at several levels of system damping.

Reference to Figure 2.34 indicates that at resonance with the driving frequency the same as the damped natural frequency, the amplitude of the motion (μ) increases, according to the damping in the system. With no damping the theoretical amplitude of the vibration is infinite.

In a system with steady state forced vibration with viscous damping the displacement of the mass is harmonic and occurs at the same frequency as the driving force *but is virtually never in phase with the external driving force.* The displacement lags the driving force by the angle α. Therefore if the driving force $F = F_o \sin \omega t$ the displacement $x = X_o \sin (\omega t - \alpha)$.

The angle α depends on the frequency ratio $r = f/f_n$ and the degree of viscous damping as illustrated in Figure 2.35. When the frequency of excitation is the damped natural frequency ($r=1$) the angle α is always 90° regardless of the degree of damping in the system.

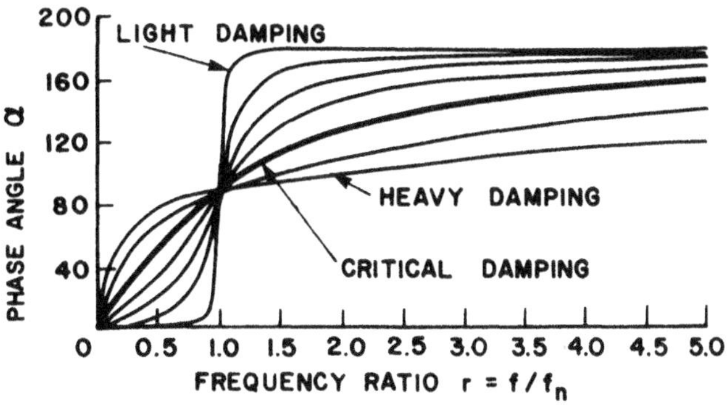

Figure 2.35 Phase angle α by which the excitation force leads the displacement in the forced vibration mode of a steady state spring-mass system as a function of the excitation frequency and at several levels of system damping.

Non-Viscous Damping

Damping is any force which resists the motion at all times. Viscous damping where the resisting force is proportional to velocity, is the easiest concept to handle mathematically. Furthermore viscous dampers are easy to make. However the viscosity of most fluids varies with temperature so that viscous damping independent of temperature is hard to achieve.

Many practical damping devices or dashpots are of the type where a fluid, liquid or gas, is squeezed through an orifice to create a resisting force. The damping force is then proportional to the *square* of the velocity of relative motion in the damper.

The frictional drag of dry sliding surfaces is virtually independent of velocity and exerts a nearly constant drag opposing the motion of a vibrating mass. This is described as *Coulomb damping*.

Hysteresis damping is nearly always present in vibrating systems with elastic restoring forces although it is rarely of sufficient magnitude to warrant consideration. It arises because of internal friction effects in any mechanical or gas spring system experiencing repeated cyclic flexing. The phenomenon is illustrated in Figure 2.36. This shows a stress/strain diagram for the metallic elements in, say, a mechanical spring. When the spring is being compressed or expanded the stress and strain both increase. When the spring is relaxed the stress and strain decrease. The curve produced during relaxation is slightly different than the curve followed during compression and expansion due to internal hysteresis effects. The area of the loop thus formed is a measure of the energy consumed by the internal friction and eventually manifested as heat.

In gas springs similar effects occur. Consider for example

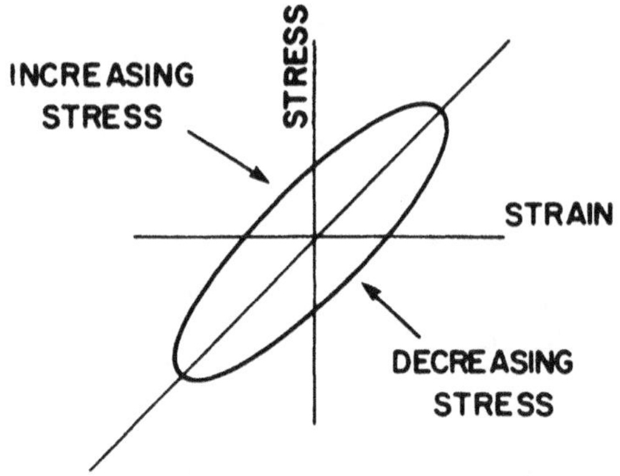

Figure 2.36 Hysteresis loop for spring element in a vibrating system.

the piston cylinder arrangement shown in Figure 2.37 where gas is contained in the space below the piston. When the mass oscillates in the cylinder gas is compressed during descent and expands during ascent of the mass. The compression and expansion follow some intermediate process between isothermal and adiabatic but as a consequence of heat transfer the curve followed by the expansion process is different to that followed by the compression process. The net effect corresponds to a *gas hysteresis* loss and can result in an appreciable consumption of energy, in fact one of principal secondary losses, of free-piston Stirling engines.

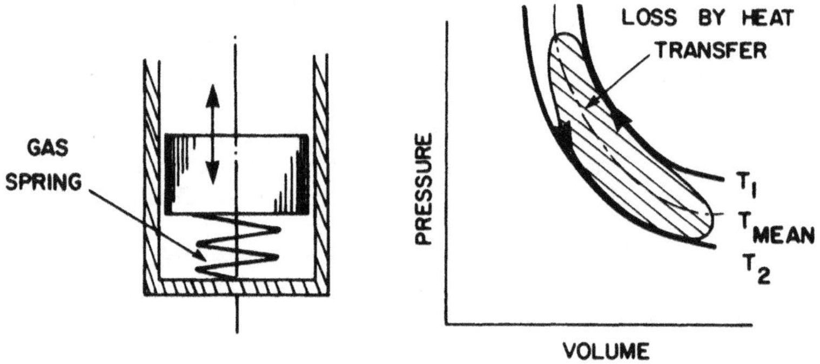

Figure 2.37 Equivalent hysteresis loop for a gas spring system resulting from internal friction and external heat transfer.

Systems With More Than One Degree of Freedom

We have dwelt at length on the simplest possible system for an elementary review of vibration phenomena. We assumed the mass was constrained to move in a vertical direction and therefore a single coordinate was necessary to completely specify the geometric location of the system mass in space. Use of the single coordinate classified the system as having a single degree of freedom.

Free-piston Stirling engines are more complicated and typically involve several masses with multiple springs and damping devices. The generalized diagram for a piston-displacer machine is shown in Figure 2.38. It comprises a system of three masses, the piston, the displacer and a cylinder with up to four springs and four

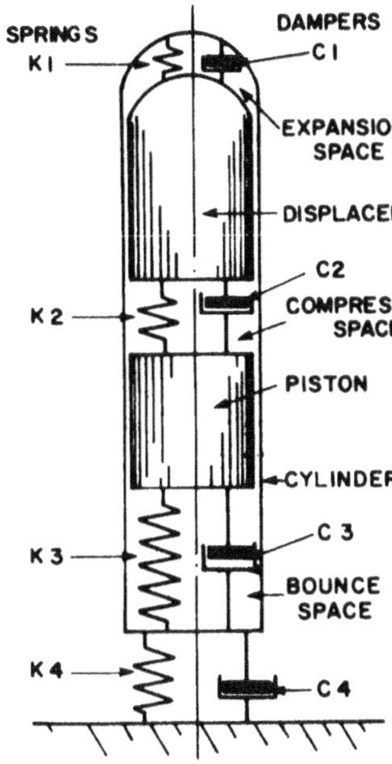

Figure 2.38 Generalized mass, spring, damper system for a piston-displacer free-piston Stirling engine.

dampers. At least three coordinates are necessary to describe the geometric location of the three masses even assuming they are constrained to move only in the vertical direction. This would then be decribed as a damped spring-mass system with three degrees of freedom. If the masses were free to rotate about their vertical axis a further three coordinates would be necessary to determine their angular location making it a system with six degrees of freedom. In practice most free-piston Stirling engines are symmetrical about the cylinder axis so that rotation in a random or controlled manner can generally be ignored for dynamic analysis.

In many cases the cylinder or piston is sufficiently massive or firmly secured to the foundations so as to be fixed in space so the problem reduces to a two degree of freedom system.

The same principles discussed above apply in systems where there is more than one degree of freedom but the details of the analysis are considerably more complicated. All the texts referenced above include closing chapters on systems with multiple degrees of freedom.

To illustrate the complexity introduced with a second degree consider the two-mass two-spring system shown in Figure 2.39. This

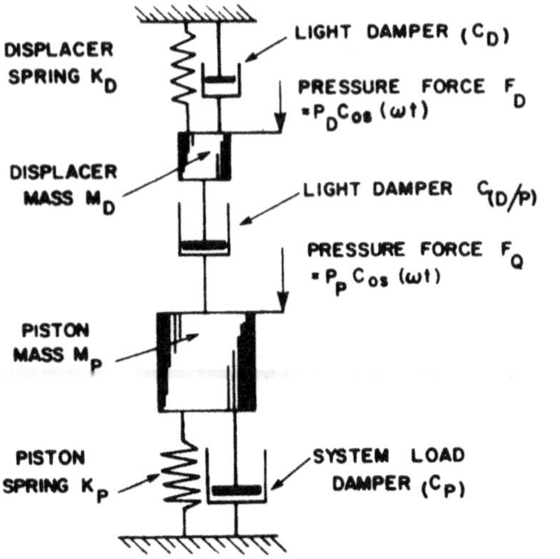

Figure 2.39 Damped spring-mass system with two degrees of freedom resembling free-piston Stirling engine.

contains two masses, a massive piston P and a much lighter displacer D with two springs of stiffness K_p and K_d. Sinusoidally varying excitation forces F_p and F_d of the same frequency and in phase but of different amplitude are applied to both P and D. This model resembles in an elementary fashion a free-piston Stirling engine having a piston and displacer in the same cylinder.

The amplitude of oscillation of the two masses as a function of the forcing frequency is illustrated in Figure 2.40. Two separate figures are drawn, one for the amplitude of the mass D and the other for the mass P. For the purpose of this figure it was assumed there was a considerable difference in the masses, i.e. P = 10D, that $K_p = 4K_d$ and that the maximum amplitude F_p and F_d of the excitation forces remained constant with $F_p = 4F_d$. In a practical free-piston Stirling engine the pressure is, of course, a function of the amplitude of oscillation and therefore the equivalent F_p and F_d would vary as the amplitude of oscillation.

Consideration of Figure 2.40 indicates that when there is no damping between the masses the system behaves as two independent, one degree of freedom systems with natural frequency of 25 Hertz for the piston and 40 Hertz for the displacer. The amplitude/frequency

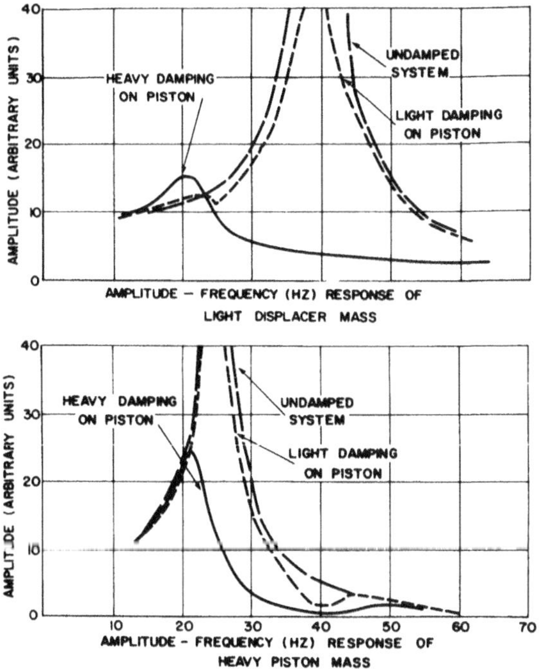

Figure 2.40 Amplitude of oscillation as a function of excitation frequency for the two masses P and D of a two-mass two degree of freedom system with damping.

characteristic is shown by the upper broken lines on the figure.

Application of moderate damping to the piston and very light damping to the displacer and with light damper coupling between the elements results in the intermediate chain dotted curves. The application of heavy damping to the piston with the same moderate damping coupling the masses and applied to the displacer results in the amplitude/frequency characteristic shown by the full lines. The resonant frequency for the piston in the damped coupled system has decreased to about 20 cycles per sec. and there is a corresponding sympathetic amplitude of vibration for the displacer at that same frequency. At the higher frequency of 40 Hertz where the displacer approaches its maximum amplitude the piston is virtually stationary.

Readers are cautioned not to read too much into Figure 2.40. It is included simply to illustrate some of the complexities of a multiple mass system resembling an idealized free-piston Stirling engine. There are really few generalities that can be made about systems with several degrees of freedom because of the great diversity of arrangements, spring coupling, damping geometry, etc. Each system must be analyzed separately. Techniques for this are given in all the major texts on mechanical vibrations.

Non-Linear Systems

Vibrating systems can be either linear or non-linear systems. In a linear system the superposition principle applies. For example if the periodic excitation force applied to a system is doubled the response of the system is also doubled.

In a non-linear system the superposition principle does not apply. It may be that the response depends on both the frequency and the amplitude of the excitation. Another characteristic feature of a linear system is that it has a singular position of equilibrium. Non-linear systems can have more than one equilibrium position depending on the conditions of equilibrium. In free-piston Stirling engines the overdriven displacer is a non-linear system (see Figure 5.5) whereas the resonant displacer, moving with simple harmonic motion, is a linear system.

Analysis of non-linear systems is very difficult and so many engineering problems are linearized to simplify analysis although some phenomena cannot be predicted by linear theory. Simple analyses of free-piston Stirling engines always assume linear systems but there are significant departures from linearity in practical engines manifest in the non-sinusoidal motion of the reciprocating elements and in the measured pressure-time characteristics.

VECTOR REPRESENTATION OF VIBRATING SYSTEMS

Introduction

Reciprocating harmonic motion can be effectively represented by the projection of the end point of a rotating vector on vertical or horizontal axes passing through the center of rotation '0' as illustrated in Figure 2.41. A vector of length X rotates counterclockwise about 0 at an angular velocity ω with a starting point when the vector is at the positive horizontal axis. The angle of rotation is then ωt where t is time. Projection of the vector measures $x = X \cos \omega t$, distance 0-a, on the horizontal axis and $y = X \sin \omega t$, distance 0-b, on the vertical axis passing through center of rotation 0. Both these projections represent simple harmonic motion but for convenience let us consider only the horizontal projection $x = X \cos \omega t$. Projection of the vector 'oc' of length X on the x axis then has the same simple harmonic motion as the displacement of a mass M on a spring in the simple one degree of freedom system shown in Figure 2.31.

Let us recall now:

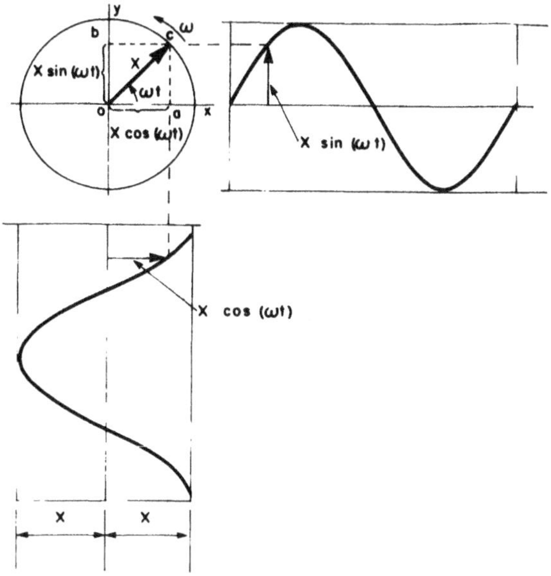

Figure 2.41 Vector representation of simple harmonic motion.

(a) the velocity of the mass M was $\dot{x} = -X\omega \sin \omega t = X\omega \cos (\omega t + \pi/2)$, and

(b) the acceleration of M was: $\ddot{x} = -X\omega^2 \cos \omega t = X\omega^2 \cos (\omega t + \pi)$.

The velocity and acceleration terms can be represented in the same way as the displacement by additional vectors of length ωX for the velocity and $\omega^2 X$ for the acceleration as shown in Figure 2.42. The velocity vector leads the displacement vector by 90° or $\pi/2$ as required by the above equation. The acceleration vector leads the displacement vector by 180° or π. Projection of the velocity and

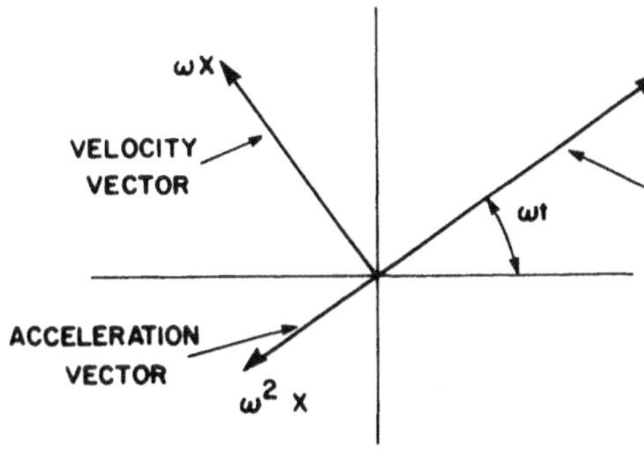

Figure 2.42 Vector representation of displacement, velocity and acceleration.

accelerataton vectors on the horizontal axis give distances from the origin 0 of $\dot{x} = X\omega \cos(\omega t + \pi/2)$ and $\ddot{x} = -X\omega^2 \cos(\omega t + \pi)$ conforming exactly with the above equations.

Vector Addition and Subtraction

Vectors representing quantities having the same *frequency* but of different *phase angle* and *magnitude* may be added or subtracted graphically as shown in Figure 2.43. In this case vector X_1 at angle

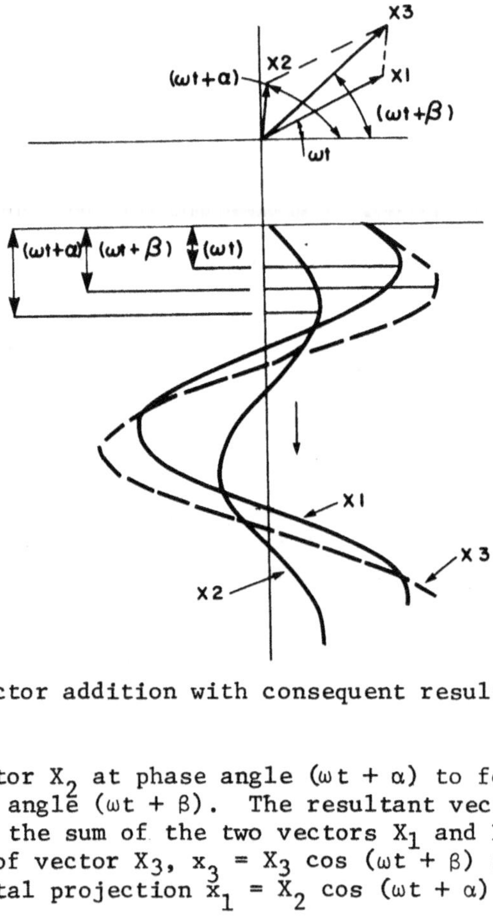

Figure 2.43 Vector addition with consequent resultant vector.

ωt is added to vector X_2 at phase angle $(\omega t + \alpha)$ to form the resultant vector X_3 at phase angle $(\omega t + \beta)$. The resultant vector X_3 then completely represents the sum of the two vectors X_1 and X_2. The horizontal projection of vector X_3, $x_3 = X_3 \cos(\omega t + \beta)$ is equal to the sum of the horizontal projection $x_1 = X_2 \cos(\omega t + \alpha)$.

Force Representation

It is convenient now to recall in a vibrating mass-spring system that the spring force Kx always *resists* the displacement and may therefore be represented by a rotating vector Kx directed opposite to the displacement vector X as shown in Figure 2.44. The damping

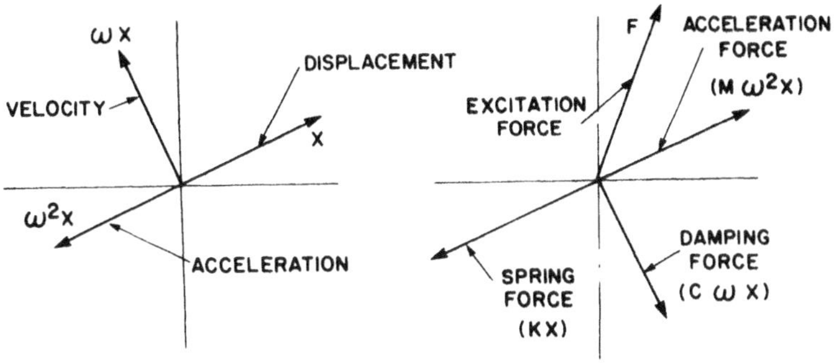

Figure 2.44 Vector representation of spring force, damping force, acceleration force and excitation force in forced vibrating damped spring-mass system.

force $C\dot{x}$ is a function of both the velocity of the mass and the coefficient of viscous damping, C, and acts always to resist the motion of the mass. It may therefore be represented by a vector of length $C\omega x$ diametrically opposed to the velocity vector. Likewise, the inertia force, $M\ddot{x}$, always resists the acceleration of the mass and may be represented by the vector $M\omega^2x$ directed in the opposite direction to the acceleration vector.

All these forces are shown in Figure 2.44(b) together with a further vector F representing the excitation force $F \cos (\omega t + \alpha)$ applied to the mass M of the simple one degree of freedom system shown in Figure 2.31. The projection on the horizontal axis of all these vectors then corresponds to the equation of motion of the mass M:

$$M\ddot{x} + C\dot{x} + Kx = F \cos (\omega t + \alpha)$$

representing the summation of the inertia force, the damping force and the spring force equated to the excitation force acting on the mass.

Vector Force Polygon

The various forces can be added and the resultant excitation force evaluated by constructing the vector force polygon shown in Figure 2.45. For convenience in this diagram the angle ωt was chosen to be zero so the displacement vector lies along the horizontal line. From such vector polygons the force F and the angle α can be determined provided X, ω, K and C are known.

For a driven system the vector representing the excitation force always leads the vector representing the displacement by an angle α. This is necessary to generate a vertical component of force to balance the damping force represented by $C\omega x$. For the case illustrated the frequency of the excitation force ω_f is less than the natural frequency ω_n and the angle α is always between 0 and 90° because the acceleration force vector $M\omega^2x$ is less than the spring force vector Kx.

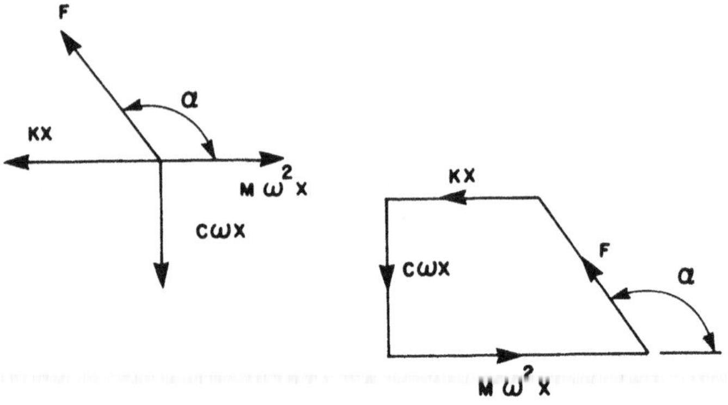

Figure 2.45 Force vector polygon for forced vibration of system with the excitation frequency less than the natural frequency.

At resonance $\omega_f = \omega_n$ and the inertia force $M\omega^2x = Kx$. Therefore the force polygon is as shown in Figure 2.46 and the angle α by which the excitation force leads the displacement is always 90°.

When the excitation frequency ω_f is greater than the natural frequency ω_n the inertia force $M\omega^2x$ exceeds the spring force Kx. The vector force polygon is as shown in Figure 2.47 and the angle α will always be between 90 and 180°.

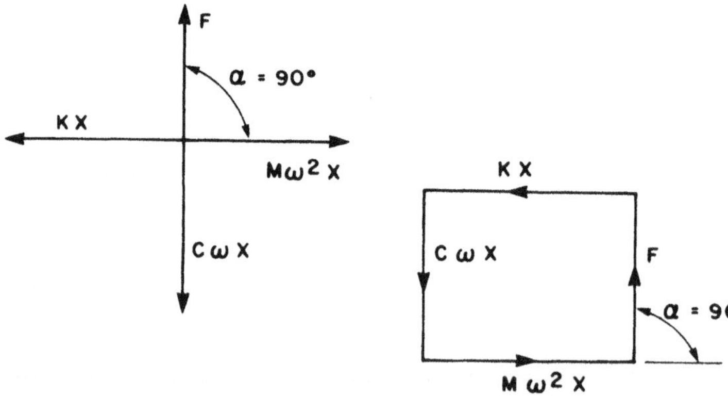

Figure 2.46 Force vector polygon for forced vibration at resonant frequency.

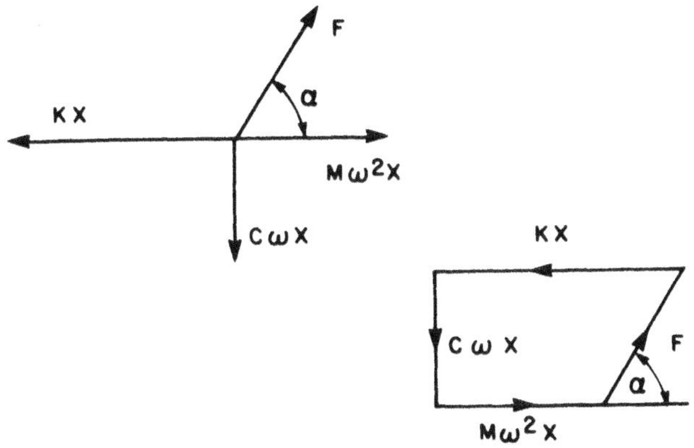

Figure 2.47 Force vector polygon for forced vibration with the excitation frequency greater than the natural frequency.

Work Input and Power Consumption

Mechanical vibration texts show the work done per cycle by the harmonic excitation force to sustain a harmonic vibration is:

$$\text{Work/Cycle} = \pi F\ X\ \text{Sin}\ \alpha$$

where F and X are the amplitudes of the excitation force and displacement respectively and α is the phase angle between them. The work input is necessary to overcome the damping frictional losses. A completely undamped system requires no input of work.

The power consumed is the work done per cycle times the frequency:

$$\text{Power} = \text{Work done/cycle} \times \text{frequency}$$

$$= \pi F\ X \sin\alpha \times (\omega/2\pi) = (\omega/2)\ F\ X \sin\alpha$$

and is a maximum at the resonant frequency.

For the simple, one degree of freedom, damped spring-mass system the power consumption at resonance can be shown to be:

$$\text{Maximum power} = F^2/2C$$

where C is the coefficient of viscous damping.

In resolution of vector force diagrams it is enlightening to consider the excitation force F as made up of two components, one, F sin α perpendicular to the displacement, and the other, F cos α parallel to the displacement. The vertical component can be regarded as the work component of the force to overcome the damper resistance.

The horizontal component corresponds to a spring force and does no work and consumes no power.

FREE-PISTON STIRLING ENGINES AS VIBRATING SYSTEMS

Free-piston Stirling engines utilize the principles of vibrating systems discussed above. The necessary requirement for their operation is the dynamic equilibrium of forces as required by Newton's Second Law, that is, the sum of all forces acting on each moving component equals the acceleration force.

In free-piston Stirling engines the principal forces are the damping forces, spring forces and the working fluid pressure forces.

The damping forces include a variety of internal mechanical and aerodynamic frictional forces as well as the resistance to motion imposed by loading devices driven by the engine. In many cases the load is attached to the piston and includes water pump plungers, the moving element of electric power generators, the cylinder of a fluid inertia pump and the piston of a gas compressor. The internal damping forces profoundly influence the dynamic characteristics of displacer motion.

The internal damping forces are likely to be negligible compared with the load damping and so have little effect on the piston motion. However, the internal damping forces are in many cases the only damping forces acting on the displacer and so have a profound influence on the dynamic characteristics of displacer motion. This in turn affects the nature of the pressure wave and hence the motion of the piston.

The spring forces acting on the piston and displacer arise from mechanical or gas springs supporting the reciprocating elements.

The gas pressure forces arise as a result of the cyclic variation of the working fluid pressure during the operation of the engine. Motion of the piston increases and decreases the total volume of the working space thereby creating a cyclic pressure variation. Motion of the displacer shuttles the working fluid between the hot expansion space and the cold compression space thereby creating a further pressure fluctuation.

This complicated situation where the oscillation of a system is sustained because of the form of its own response is described as a 'limit cycle' and is discussed in the standard texts on engineering control systems. A pictorial representation of a limit cycle is given in Figure 2.48. The full line shows the stable cyclic relationship of displacement x and the velocity $\dot{x}$ of a given system. The response following any arbitrary starting condition, point A or C, will eventually descend into the stable cyclic relationship (at B and D). However, there may also be other operating regimes

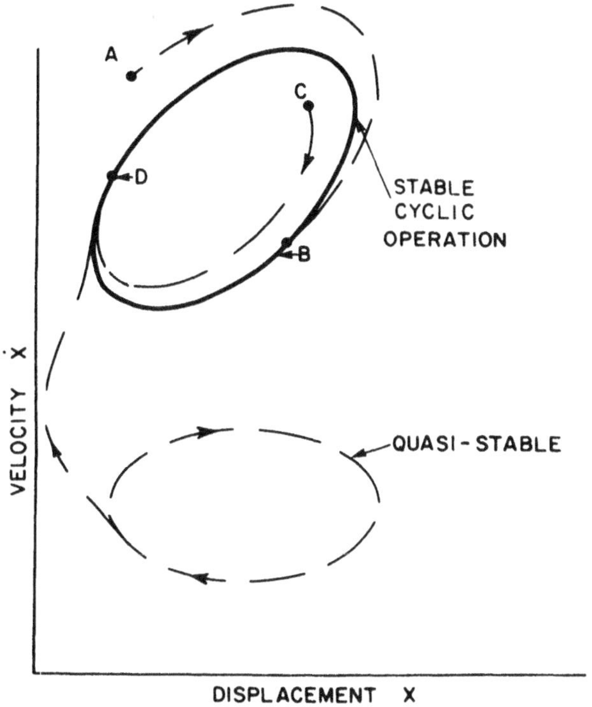

Figure 2.48 Coupled displacement-response characteristic of a limit cycle.

for the system. One is indicated in the lower figure and if the system is started anywhere along its path the system will remain locked onto the path. Any subsequent random perturbation causing the system to be dislodged from the quasi-stable cyclic path will result in an excitation/response characteristic which will eventually move the system into the stable cycle relationship. This inevitable descent to the stable cyclic state is explained by Hamilton's Principle of Minimum Energy States; a concept beyond our discussion here but found in books on engineering control systems and advanced dynamics.

The important implication of this for free-piston Stirling engines is that an engine has a preferred stable operating frequency which, in fact, is close to the resonant frequency of the largest dynamic mass involved in the engine. Beale *et al* (1973) suggest the operating frequency may be calculated as:

$$f_n = c\sqrt{K/M}$$

where c = constant usually in the range 0.7 to 1.0,
K = spring stiffness,
M = piston mass.

This will be discussed in more detail below.

VECTOR REPRESENTATION OF FREE-PISTON STIRLING ENGINES

To investigate forces involved on the reciprocating elements of free-piston Stirling engines it is convenient to use the method of vector representation outlined above for vibrating systems.

a) Piston/Displacer System, Both Elements Sprung to Ground

As the first example of vector representation of free-piston Stirling engines consider the familiar arrangement of the Beale piston-displacer free-piston Stirling engine with a gas spring in the bounce space. The elements of the engine are shown in Figure 2.49. For convenience the piston and displacer are represented as coupled to independent mechanical springs but the springs are most often gaseous springs. Three damping devices are shown, C1, C2 and C3. The

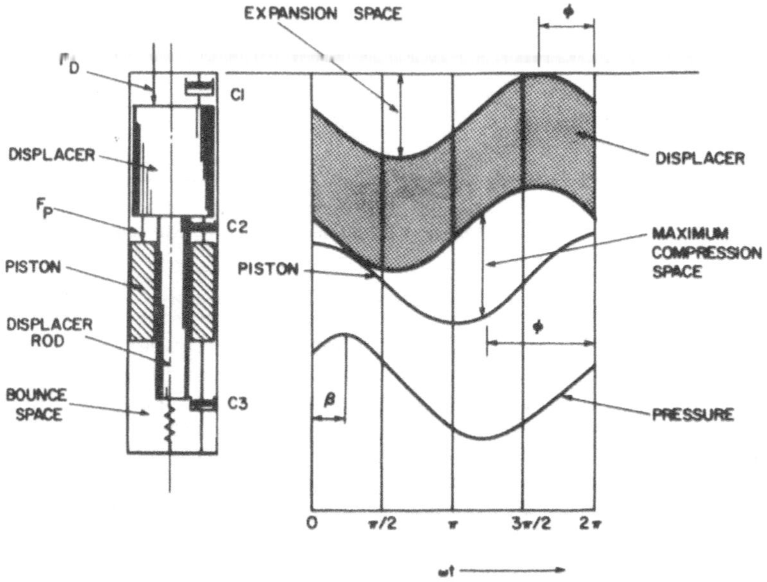

Figure 2.49 Free-piston Stirling engine of the piston-displacer in a single cylinder with a common spring in the bounce space.

major damping device C3 couples the piston to ground and is the means whereby work is extracted from the piston. The other two damping devices are of a minor nature representing viscous friction losses (from both mechanical and aerodynamic sources) of the displacer moving with respect to the piston and the displacer moving with respect to ground.

Typical time displacement and pressure time diagrams for the piston and displacer are also shown in Figure 2.49. For convenience we take the top dead centre (TDC) position of the piston as the zero datum. The displacer is at its top dead centre position some time earlier, shown on Figure 2.49 as the angle ϕ. The pressure curve attains a maximum value angle β *after* the piston is at TDC.

We can represent these cyclic displacements and pressures by vectors as shown in Figure 2.50(a). It is customary practice to

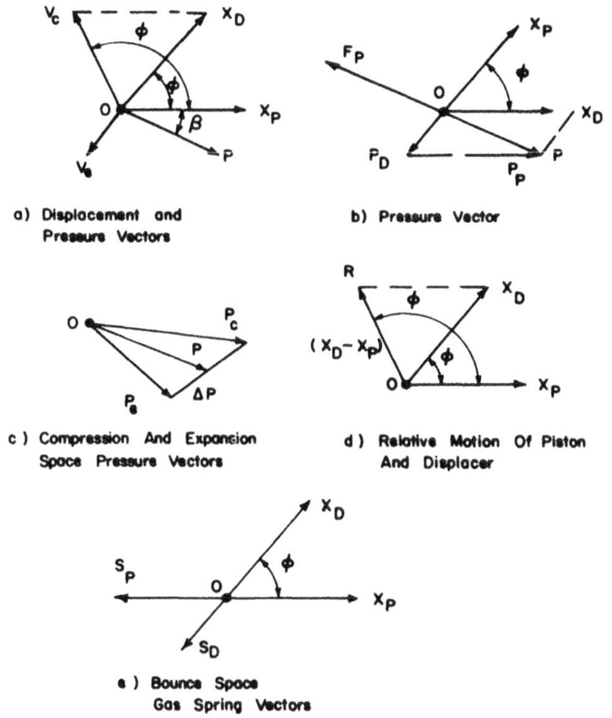

Figure 2.50 Vector representation for the piston displacer free-piston Stirling engine with common spring to ground in the bounce space.

represent the reference displacement vector horizontally pointing to the right at the datum $\omega t = 0$. Here we have chosen the piston displacement as the reference displacement and have represented this by the horizontal vector X_p. The displacer vector leads the piston vector by angle ϕ and is represented by vector X_d. We adopt the convention that '0' is the mid-stroke position for both the piston and displacer and that motion from the mid-point towards the TDC is positive and represented by vectors drawn to the right of the mid-point.

Motion of the displacer from the mid-point towards the

TDC position results in a reduction of the expansion space (above the displacer). Therefore the instantaneous value of the expansion space volume, V_e, may be represented on Figure 2.50(a) by a vector, V_e, in line but 180° out of phase with the displacer vector, X_D.

The instantaneous value of the compression space depends on the motion of both the piston and displacer. Motion of the displacer from the mid-point towards the TDC position increases the compression space (below the displacer) but a similar motion of the piston (from the mid-point towards the TDC position) decreases the compression space. Therefore the instantaneous volume of the compression space (above the piston/below the displacer) varies as the *difference* in the displacements of the piston and displacer.

Vector *subtraction* of the piston motion X_p from the displacer motion X_d gives the resultant vector V_c for the instantaneous volume of the compression space. Maximum compression space volume is attained angle θ ahead of the piston TDC position.

To understand the relationship of pressure and displacement better it is convenient to once again consider the piston and displacer at the mid-point of their strokes.

Now, holding the piston stationary, move the displacer a short distance towards the TDC position. This displaces fluid from the hot expansion space to the cold compression space resulting in a small *decrease* in the pressure. We can represent this in Figure 2.50(b) by the short vector, P_D, acting in opposition to the displacement vector, X_D. There is a further minor effect on the pressure resulting from the displacer motion. The volume variations above and below the displacer resulting from the displacer motion differ by the volume of the displacer rod. As the displacer moves upward towards the TDC position more displacer rod enters the compression space above the piston tending to *increase* the pressure. This can be represented by a vector P_R acting in concert with the displacement vector, X_D. The effect is so small it can frequently be ignored.

Motion of the piston from the mid-stroke position towards the piston TDC with the displacer held stationary at mid-stroke causes an increase in the pressure. This can be represented as in Figure 2.50(b) by the vector, P_P, acting in the same direction as the piston.

Addition of the displacer pressure vector, P_D, and the piston pressure vector, P_P, produces the resultant vector P trailing the piston displacement, X_D, by angle β as shown in Figure 2.50(b). The convention adopted to represent the force on the piston and displacer due to pressure in the working space is that an *increase* in pressure results in a *negative* force on the piston and displacer. Therefore the force due to pressure may be represented by the vector, F_P, in Figure 2.50(b) drawn in opposition to the pressure vector P.

In a practical engine a pressure drop will occur across the regenerator, heater and cooler producing a phase lag between pressure changes in the compression and expansion spaces. This is illustrated in Figure 2.50(c), where Δp represents the pressure drop applied to the ideal pressure vector P, resulting in two pressure vectors, $P_{(c)}$ representing the pressure variation in the compression space and leading the pressure vector, $P_{(e)}$, representing the pressure variation in the expansion space.

Damping forces arise from the fluid damper C2 shown in Figure 2.49. In practice this is principally the fluid seal along the displacer rod passing through the piston. To resolve these damping forces we need to consider the *relative* motion of the piston and displacer. This is obtained by subtracting the vectors X_P and X_D representing the piston and displacer motions as shown in Figure 2.50(d). The relative motion is represented by vector R at angle θ.

Consider now the forces acting on the piston and displacer as a result of the gas pressures in the bounce space. These are, in fact, the spring forces, S_P and S_D, respectively, arising from the bounce space gas springs. We know from the earlier discussion that spring forces act always to oppose the motion of a reciprocating mass. Therefore, we can represent the bounce space gas spring forces by vectors S_P and S_D drawn in opposition to the displacement vectors X_P and X_D for the piston and displacer, respectively, as shown on Figure 2.50(e).

Now we are in a position to draw the force polygons representing all the forces acting on the piston and displacer. At this stage it is convenient to recall that the pressure forces in the working space acting on both the piston and displacer are generated by a combination of the cyclic motion of the piston and displacer and the energy flow *into* the expansion at high temperature and *out of* the compression space at low temperature. The pressure forces are therefore resultant forces arising from both the motion and the energy flow. We retain the convention that an *increase* of pressure in the working space results in a *negative* force on the piston and displacer.

The force polygons for the piston and displacer are drawn in Figure 2.51. Figure 2.51(a) is simply a restatement of Figure 2.50. Figure 2.51(b) is the force polygon for the piston*. The force F·(S) acting along a-b is the spring force on the piston acting in opposition to the piston displacement. Force F(C3), acting along b-c, is the damping force arising from the damping device C3

*At this point readers may find it convenient to refer to Figure 2.4 showing the representation of spring, damping and acceleration forces.

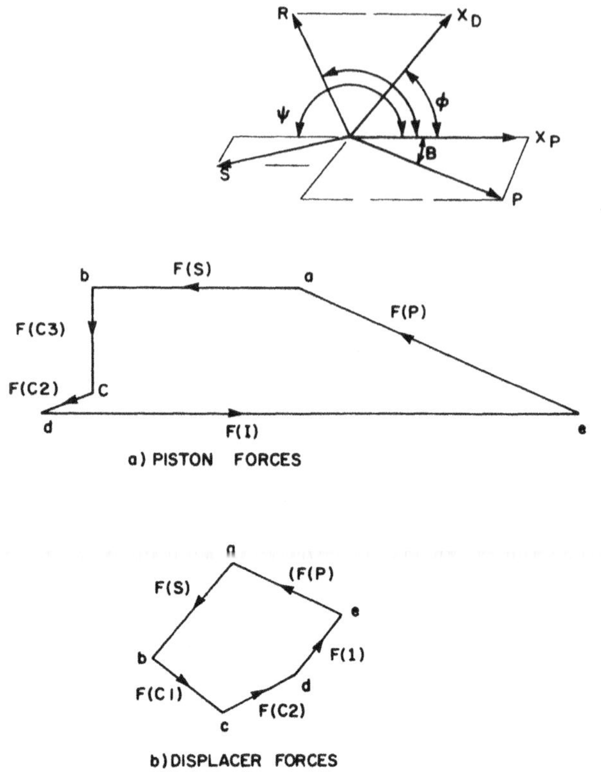

Figure 2.51 Vector force polygons for the piston and displacer.

absorbing the work output of the piston. The damping force acts in opposition to the *velocity* vector which leads the *displacement* vector by 90°. Hence the damping force lags the displacement vector by 90° and is drawn perpendicular to the displacement vector. The damping force F(C2), acting along c-d, arises from the viscous damping coupling of the piston and displacer, arising because of the gas spring coupling the piston and displacer and because of aerodynamic pumping losses in the heat exchangers of the engine. It acts perpendicular to the vector OR representing the relative motion of the piston and displacer.

The inertia force F(I), acting along d-e, arises from the *acceleration* vector which leads the displacement vector by 180°. The inertia force acts in opposition to the acceleration vector and therefore acts along the line of the displacement vector.

The force F(P), acting along e-a, is the excitation force on the piston due to pressure and acts along the line, but in the opposed direction, to the pressure vector P.

This pressure force must always close the free polygon for steady-state operation of the engine. If the force polygon does not close it means the engine cannot operate at that condition and so the *stroke* and *phasing* of the piston and displacer will be automatically adjusted. This causes a readjustment of the pressure phasing and magnitude, the spring force, the damping force and the inertia force to ensure, by Newton's Second Law, a dynamic balance of all the forces acting, and, consequent closure of the force polygon.

The force polygon for the displacer forces can be drawn in similar fashion. The forces involved are much smaller than those on the piston for the mass of the displacer is much less and the pressure forces act only on the displacer rod area. For clarity the displacer force polygon has been drawn to a larger scale than the piston force polygon. The spring force F(S) acts along a-b, in opposition to vector X_d in Figure 2.51(a). The light damping force F(C1) acts along b-c perpendicular to the line of action of the displacer displacement vector X_d. The light damping force F(C2) acts along c-d, perpendicular to the relative motion vector R in Figure 2.51(a). The inertia force d-e acts along the line of action of the displacement vector X_d.

The resultant pressure force vector F(P) acts along 'e-a' parallel but opposed to the pressure vector P.

The magnitude of the forces acting are:

a) <u>On the piston:</u>

$$F(S) = K\,x_p$$

where K is the spring stiffness,
$K = \gamma A_B^2\, P/V$,
x_p = piston displacement,
γ = ratio of specific heats of bounce space gas,
A_B = area of bounce space cylinder,
P_b = average gas pressure in bounce space,
V = volume of the gas spring bounce space at the mid-point of piston stroke.

$$F(C3) = C_3\,\dot{x}_p = C_3\,\omega X_p$$

where C_3 = coefficient of viscous damping of the viscous damper C3,
ω = $2\pi f$,
f = frequency of operation.

$$F(C2) = C_2\,\dot{x}_R = C_2\,\omega R$$

where C_2 = coefficient of viscous damping of the damping between the piston and displacer because of gas spring or heat exchanger losses,
$\dot{x}_R$ = relative velocity of the piston and displacer,

R = relative displacement of the piston and displacer.

$$F(I) = M_p \ddot{x}_p = M_p \omega^2 X_p$$

where M_p = mass of the piston,
$\ddot{x}_p$ = acceleration of piston.

$$F(P) = (A_c - A_R)\ P$$

where A_c = area of cylinder,
A_R = area of displacer rod,
P = working space pressure (assumed to be the same throughout the working space).

b) <u>On the displacer:</u>

$$F(S) = KX_d \cos(\psi - \phi)$$

where K = spring stiffness defined above,
x_d = displacer displacement,
ϕ = angle by which the displacer motion leads the piston.

$$F(C1) = C_1 \dot{x}_d = C_1 \omega X_d$$

where C_1 = coefficient of viscous damping of the damper C1 coupling the displacer to the cylinder (ground) arising from gas spring and heat exchanger pumping losses.

$$F(C2) = C_2 \dot{x}_R = C_2 \omega R$$

defined as above.

$$FI = M_d \ddot{x}_d = M_d \omega^2 X_d$$

where M_d = mass of displacer,
$\ddot{x}_d$ = acceleration of displacer.

$$F(P) = A_R P$$

defined as above.

A feature of this configuration is the relatively small inertia force component available to the displacer vector force polygon (force F(I) along (d-e)) in Figure 2.51(c). This is the reason why engines of this type must be constructed with very light displacers compared with other varieties considered below.

We have dwelt in painstaking detail with the above case

principally for the benefit of readers unfamiliar with the techniques of vector representation and analysis. Now in the following we shall apply the same techniques to other configurations of free-piston Stirling engines. It will be left largely as an exercise for readers to follow through the construction of the force polygons in detail.

b) Piston/Displacer System with the Displacer Sprung to Ground

The dynamic constituents of this configuration are shown diagrammatically in Figure 2.52(a). The displacer is made up of two

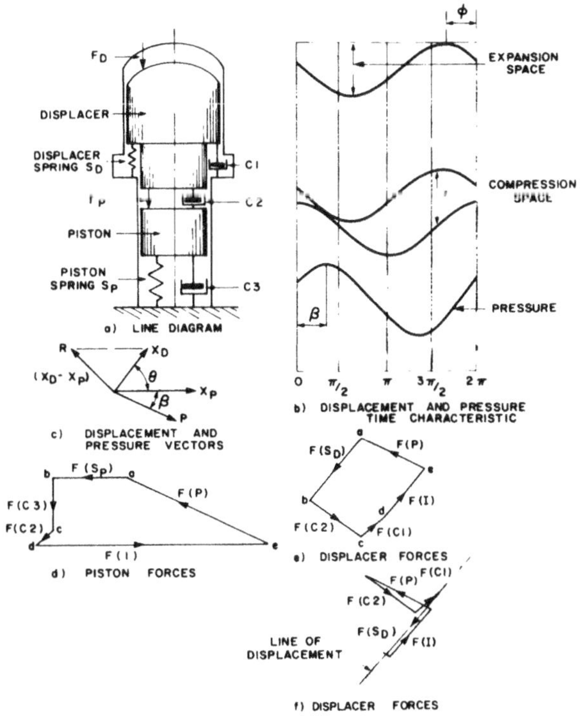

Figure 2.52 Piston-displacer machine with the displacer sprung to ground.

cylinders of different diameters and in tandem operating in a cylinder with different diameters. The concentric annulus of cylinder around the lower part of the displacer constitutes the displacer gas spring. This may or may not be enlarged as shown depending on the spring characteristics required. There is also viscous damping coupling, C1, between the displacer and the cylinder which simulates gas spring hysteresis losses.

The piston oscillates in the lower, smaller, cylinder with

gas spring S_P and viscous damper (the engine load) C3 coupling to the cylinder. There may or may not be a viscous damping coupling between the displacer and the piston, C2.

The displacement and working space pressure time diagram for this configuration are shown in Figure 2.52(b) and the equivalent vector representation in (c). The vector force polygons for the piston and displacer are shown in Figure 2.52(d) and (e) respectively.

This configuration can provide a strong displacer drive and is capable of operating engines at higher frequencies with relatively massive displacers.

Further insight and interpretation of the force polygons can be gained by constructing the polygon for the displacer as shown in Figure 2.52(f) rather than diagram (e). The sum of the damping forces F(C1) and F(C2) and the pressure force F(P) brings the force diagram onto the line of displacement Y-Y. A spring of sufficient strength is necessary to provide the resultant force for resonance force $F(S_D)$, acting in opposition to the displacer inertia force F(I) along the line of displacement.

c) <u>Piston/Displacer System with the Displacer Sprung to the Piston</u>

This piston-displacer engine configuration is shown in two alternative arrangements in Figure 2.53(a). The displacer gas spring is contained in a cavity within the displacer or in the piston as shown. There is the customary spring/damping coupling of the piston to ground, minor damping of the displacer and minor damping between the piston and displacer.

Figure 2.53(b) shows the displacement and working space pressure-time diagrams for the configuration and diagram (c), the equivalent vector representation. The vector force polygons for the piston and displacer are shown in diagrams (c) and (d).

In this arrangement neither the pressure or the spring force are parallel with the displacer line of displacement. As a consequence there is a relatively large flow of work from the displacer to the piston. This leads to a high spring constant compared with the alternative displacer sprung to ground. The displacer spring hysteresis tends to be high and as a consequence the overall thermal efficiency tends to be lower. Moreover, the displacer rod area must be considerably larger than in cases where the displacer is sprung to ground (Senft (1979)).

Despite these drawbacks the displacer sprung to piston configuration is generally simpler mechanically than when the displacer is sprung to ground and was used frequently in small early free-piston Stirling engines.

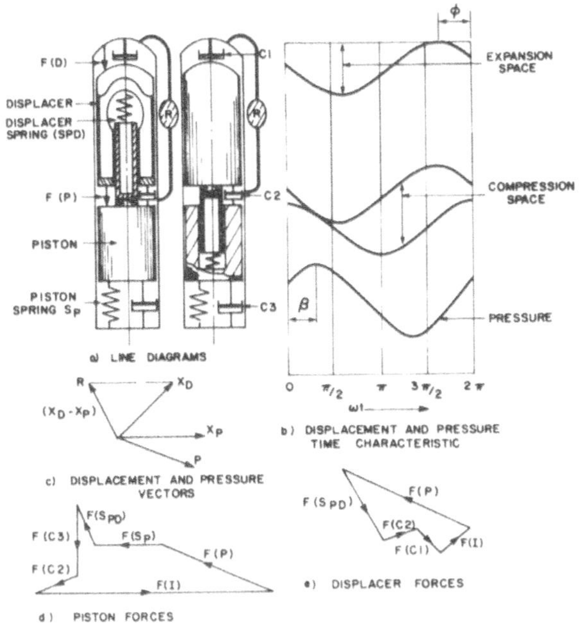

Figure 2.53 Piston-displacer machine with the displacer sprung to piston.

d) Split-Stirling Engine

This configuration is shown in Figure 2.54(a). The resonant piston P oscillates in a cylinder with a gas spring and heavy viscous damping (the engine load). The displacer oscillates in a separate cylinder and is sprung to ground with a minor damping device C1 also coupled to ground. There is some viscous coupling C2 between the piston and displacer arising from frictional pumping losses in the connecting ducts and heat exchangers. The displacement, and pressure-time diagrams are shown in Figure 2.53(b), the equivalent vector representation in diagram (c) and the piston and displacer vector force polygons in diagrams (d) and (e).

e) Two-Piston Stirling Engine

An arrangement for a two-piston Stirling engine with two parallel vertical cylinders coupled at the upper end is shown in Figure 2.55(a).

In the ideal case piston 1 would be a massless damper and piston 2 an undamped resonant mass. The pressure vector would be in phase with the displacement vector of piston 2 and at right angles to the displacement vector for piston 1 (Figure 2.55(b)).

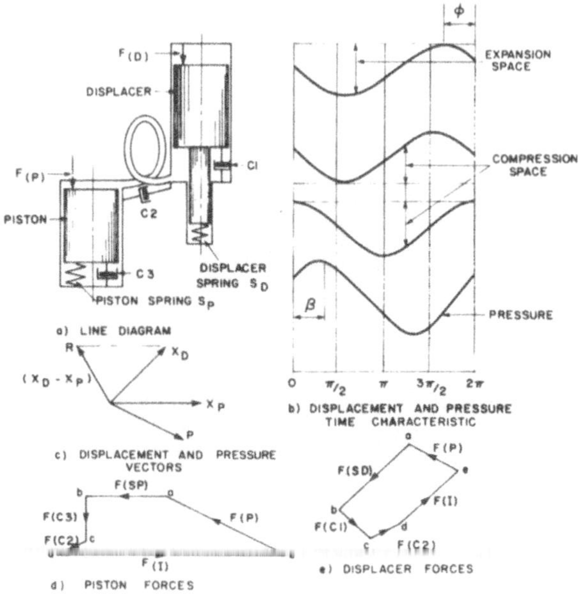

Figure 2.54 Split-Stirling engine.

In such a case piston 1 experiences a pure damping force and piston 2 experiences a pure spring. The frequency of operation would depend on the mass of spring 2 and the system spring constant, the sum of the two gas springs in the working space and the bounce space.

Unfortunately, it is not possible to devise an actual engine with these ideal characteristics. The more practical case is shown in Figure 2.55(c). The displacement vector X1 for piston 1 leads the displacement vector X2 for piston 2. The pressure vector, P_W, for the pressure of fluid in the working space will be somewhere between the displacement vectors X1 and X2. This is because displacement of either piston from the mid-stroke position with the other held stationary will result in an increase in the pressure. However, upward motion of piston 1 will displace fluid from the hot expansion space to the cold compression space and so moderate the pressure increase. Upward motion of piston 2 displaces fluid from the cold to the hot space and so enhances the pressure increase. Therefore the pressure vector P will lie closer to the displacement vector X2 than to displacement vector X1.

The pressure vector P_B for the bounce space is drawn opposite the bisecter of the (displacement) vectors X1 and X2 for there are no temperature effects to be considered in this case.

The corresponding force diagrams are drawn in Figures 2.55(d) and (e).

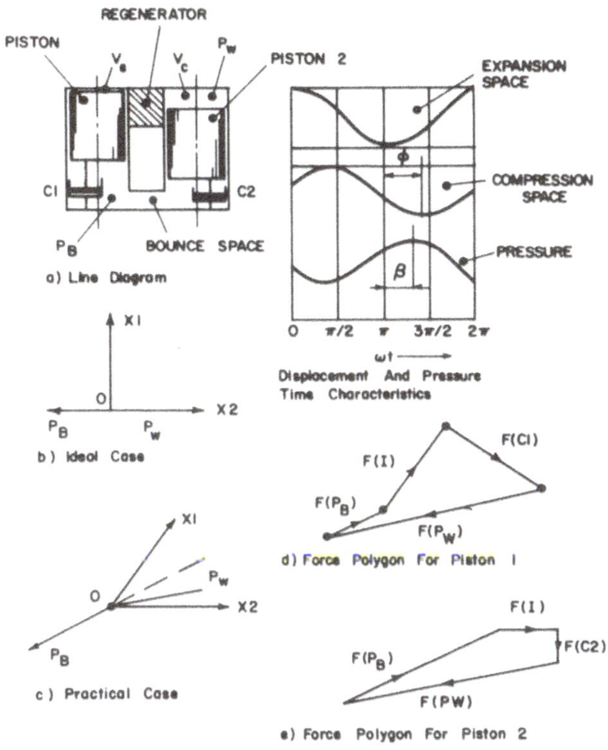

Figure 2.55 Two-piston Stirling engine.

f) Free-Cylinder Stirling Engine

A free-cylinder Stirling engine is illustrated in Figure 2.56. There are many possible variations but for the purpose of this discussion we assume the arrangement of a piston-displacer in a single cylinder. We further assume the piston so massive compared with the cylinder and displacer as to remain motionless and so represent the virtual ground. In that case the displacer and cylinder oscillate cyclically as shown in the displacement-time diagram. The corresponding pressure curve is included.

Work is taken from the system through the heavy damping device C3. There is light damper coupling (C2) of the displacer to the piston (ground) and between the displacer and the cylinder C1. The displacer spring S_D is contained within the piston. There is a further spring connection S_C between the cylinder and ground (shown here as the piston).

The vector displacement and pressure diagram is shown in Figure 2.56(c) and the vector force diagram for the cylinder and displacer in diagrams (d) and (e) respectively.

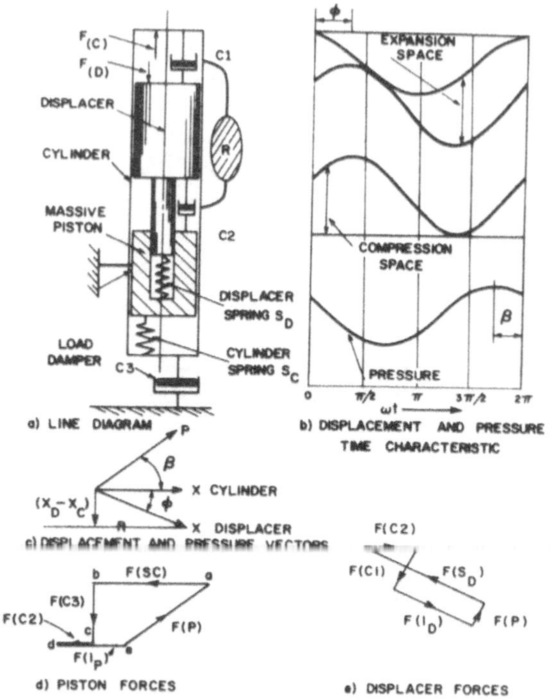

Figure 2.56 The free-cylinder Stirling engine.

g) Double-Acting Stirling Engine

A double-acting Stirling engine arrangement is illustrated in Figure 2.57. We have chosen to show a three cylinder arrangement but it could just as easily be a four, five or six cylinder machine.

The upper expansion space of one cylinder is coupled through heat exchangers including a regenerator, to the lower compression space of the adjacent cylinder to form three separate Stirling systems: i.e. $V_{E1} + V_{C2}$, $V_{E2} + V_{C3}$ and $V_{E3} + V_{E1}$. The reciprocating elements, piston-displacers, move with simple harmonic motion but are mutually 120° out of phase. The displacement-time diagrams are shown in Figure 2.57(b). The piston-displacer is made up of two concentric cylinders in tandem and operates in a cylinder having concentric bores to suit the two diameters of the piston displacer. The annulus below the displacer at the shape of section is the displacer gas spring space. It provides a gas spring of sufficient stiffness to oscillate the resonant mass of the piston-displacer. Every mass is coupled to a heavy damping device C producing the output work of the engine. There is also a light damper connection between the adjacent reciprocating masses because of aerodynamic friction losses.

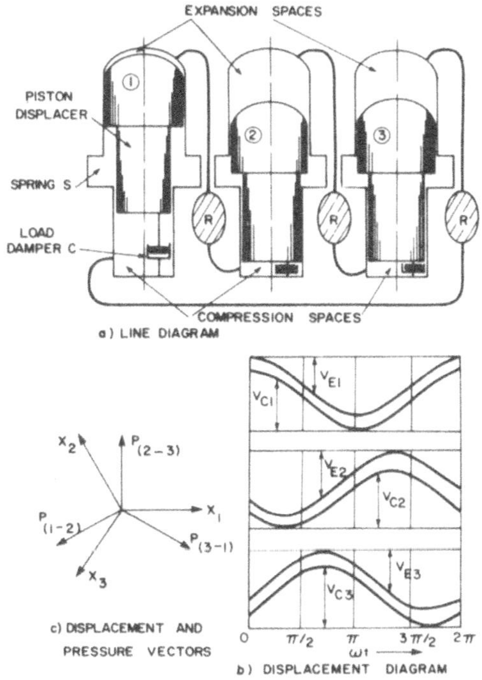

Figure 2.57 Three cylinder Siemens Stirling engine.

Referring now to the Stirling system spanning the adjacent cylinders 1 and 2. The displacement-time diagram for these two elements are superimposed in Figure 2.58. A possible pressure-time curve for the system is also shown. The corresponding displacement and pressure vectors for the system are sketched in Figure 2.58(b) as vector X1, vector X2 for the displacements and vector P1-2 for the pressure. The remaining vectors X3 and pressure P2-3 and pressure P3-1 are added by symmetry. The vector force diagram for the forces acting on the piston-displacer 1 is given in Figure 2.58(c). The forces are the pressure force due to the pressure above the displacer F(P1-3), the spring force F(S), the damping force F(C1), the light damping force F(C1-2), the pressure force due to pressure below the displacer F(P1-3) and the inertia force F(I1). The force polygons for the other two systems are of course identical.

It will be clear from the above that vector representation is a powerful method for elementary analysis and interpretation of Stirling cycle systems and is particularly appropriate for free-piston Stirling engines in resonant operation. Fokker (1978) and Ackerman (1981) have discussed the use of vector representation of the Stirling cycle. The method is used routinely by Cooke-Yarborough and colleagues at the Atomic Energy Research Establishment in their work on the thermo-mechanical generator (see Chapter 6) and is widely used at Sunpower and other establishments concerned with free-piston Stirling engine development.

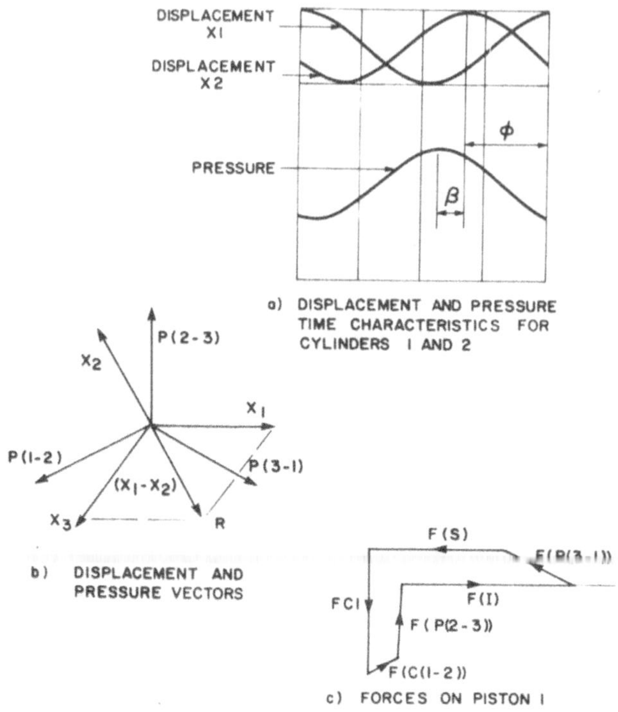

Figure 2.58 Vector representation of double-acting Siemens Stirling engine.

PART III - SOME PRACTICAL CONSIDERATIONS

INTRODUCTION

Free-piston Stirling engines involve few moving parts and appear to be quite simple and straightforward. This is deceptive. Although the machines are mechanically simple their proper design and operation requires substantial engineering expertise and very high levels of craftsmanship. Important practical considerations are discussed below.

PISTON CENTERING

No completely adequate seal has yet been devised to contain fluid in the working space yet remain sufficiently flexible for unrestrained operation of the piston.

Some leakage of fluid between the working space and the

bounce space is therefore inevitable. In the ideal situation, shown in Figure 2.59(a), the working space pressure variation is represented as purely sinusoidal with the bounce space pressure constant. Symmetry of the pressure differences suggests that leakage from the working space to the bounce space during the half cycle A-B would be compensated by the exactly equivalent leakage in the reverse direction during the remaining half cycle B-C. Even here, however, some net leakage would occur for the mean pressure on one half cycle is greater than the mean pressure during the succeeding half cycle. Furthermore, the slightest mechanical imperfections in the seal or machine surfaces, i.e. taper or varying eccentricity, will increase the flow. A preferential leakage rate one way or the other will always be found in the most carefully made seals.

In practical engines the pressure curve is never completely sinuosidal but frequently has the 'peaky' characteristic shown, exaggerated, in Figure 2.59(b). The volumetric flow through an aperture is proportional to the difference of the *squares* of the pressures acting, i.e. $V = (p_1^2 - p_2^2)$. Therefore as a consequence of the peaky pressure curve there is a tendency for fluid to leak *out* of the working space and as a consequence for the piston to 'creep' into the hot end of the cylinder.

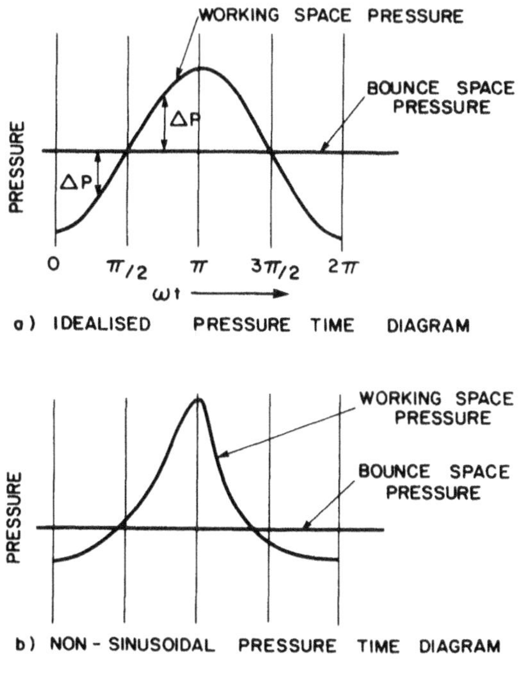

Figure 2.59 Fluid leakage potential between the bounce space and working space.

The piston can be restrained to its proper zone of oscillation by various methods which induce a compensating leak of fluid when the piston moves beyond prescribed limits. Two methods devised by Sunpower Inc. and reported by Wood (1980) involving a controlled leak between the working space and bounce space are illusrated in Figure 2.60. Another method involving compensatory leakage, but considerably more complicated, is the 'store and dump' technique illustrated in Figure 2.61.

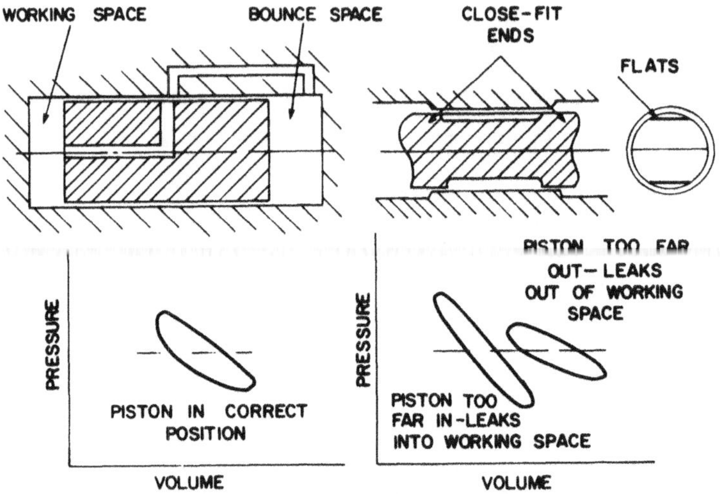

Figure 2.60 Piston centering by controlled leaks.

Even more complex is the active method for centering pistons shown in Figure 2.62. This involves precise control of the compensating leak by means of a microprocessor controlled fluid valve. Such complication is unsuited for small engines but are attractive for large engines where the piston may be required to operate in one of several modes depending on the load conditions. This could entail adjustment of the piston position or spring stiffness to change the power output. Microprocessor control is well suited to large sophisticated engines of this nature.

In piston-displacer machines positive centering of the displacer is not always necessary, particularly in smaller engines. For such cases it is adequate to allow the displacer completely free motion between given 'stops' with perhaps a fluidic damping device incorporated to minimize impact shocks. One form of double-ended fluidic retarder with rubber 'O' ring impact absorbers is shown in Figure 2.63.

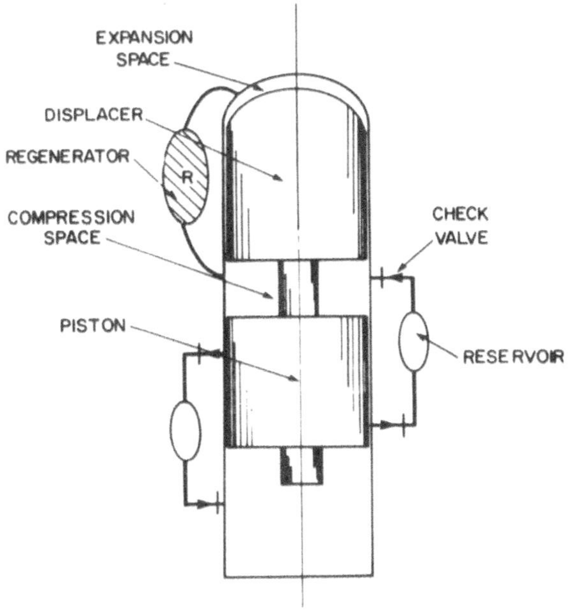

Figure 2.61 The store and dump technique for piston centering.

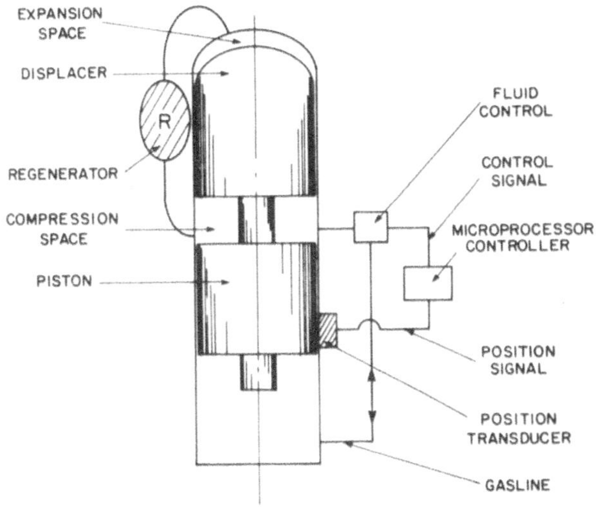

Figure 2.62 Active method for piston centering.

SEALS

Seals are the perennial problem of Stirling engines. Things are easier in free-piston Stirling engines than in machines with a

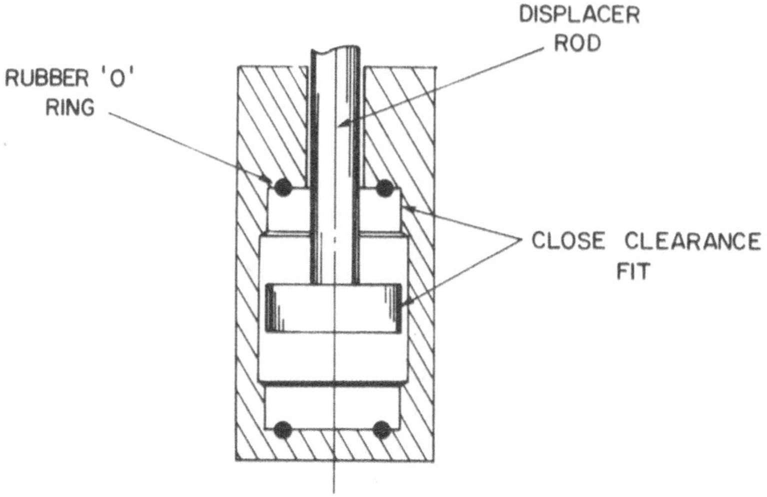

Figure 2.63 Double-ended fluidic retarded with rubber 'O' ring impact absorbers.

kinematic mechanism necessitating oil lubrication for in these cases the seal has to prevent oil entering the working space as well as containing the fluid. Oil in the working space contaminates, and eventually blocks, the regenerator flow-ways.

Free-piston Stirling engines do not have the heavy piston side forces and other demanding bearing requirements of kinematic engines. As a consequence there is no need for oil lubrication and the working fluid itself can be used as the lubricant.

Gas lubricated bearings are characteristically high precision components with very small clearances so their application in free-piston Stirling engines limits the use of conventional rubbing seals of carbon or teflon base. The detritus inevitable generated by rubbing seals catastrophically affects the free action of gas lubricated bearings. Fortunately the use of gas lubricated bearings with their consequent high precision does permit the concomitant use of very close tolerance low friction seals. Excellent results have been obtained in practice. Typical designs for close tolerance seals are given in Figure 2.64.

Grinnel (1956) has shown that the rate of leakage of gas in close tolerance seals:

$$\dot{m} \propto \frac{h^3}{L}(p_1^2 - p_2^2)$$

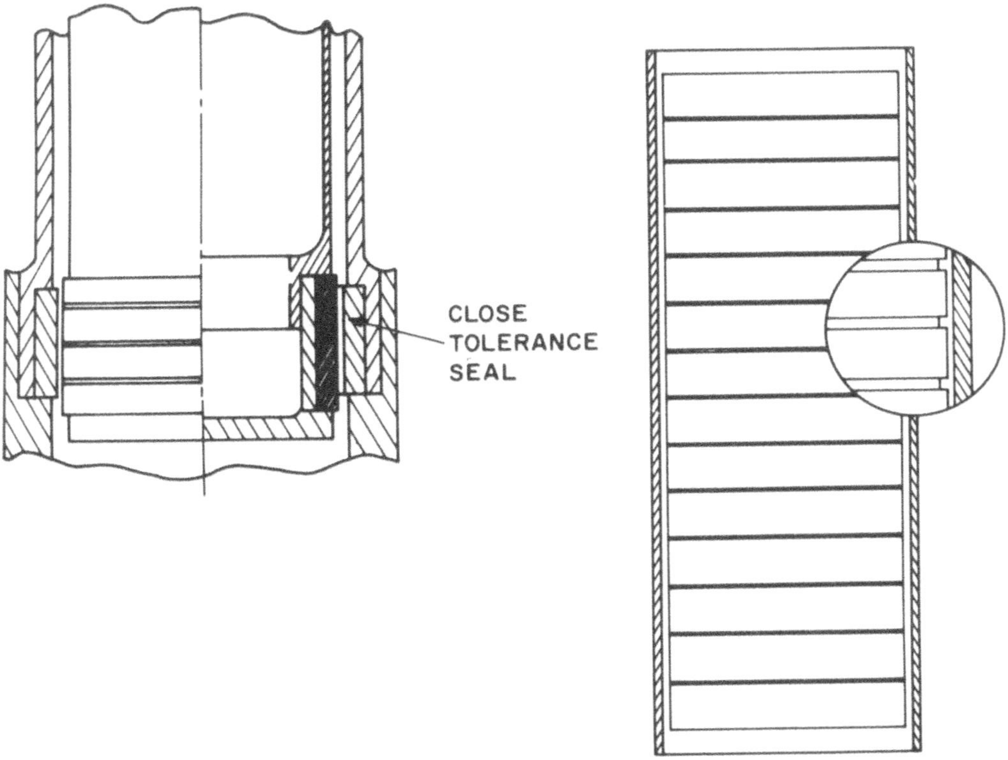

Figure 2.64 Typical close tolerance seals for free-piston Stirling engines.

where $\dot{m}$ = mass rate of leakage,
h = clearance,
L = length of seal,
p_1 and p_2 are the fluid pressures across the seal.

This equation shows the clearance 'h' is the critical dimension of a close tolerance seal and must be reduced to the minimum value achievable. Frequently close tolerance seals are ground and lapped. They can be made as individually matched pairs to high precision at surprisingly moderate cost and then incorporated in the engine with due care to avoid distortion from mechanical or thermal stressing. It is good practice to isolate the seal elements from any load carrying or structural function.

Hard anodized aluminum is a satisfactory material for low clearance seals. Aluminum is attractive because of its low mass and the hard anodizing process endows it with a very hard surface. The seal must initially be prepared with a mutual clearance in the seal elements of about 1/1000 of the diameter plus an allowance of about 0.002 inch for each anodized surface. Swelling occurs during the anodizing process and is sufficient to permit minor grinding to establish final concentricity. The seal pair may then be completed by honing or lapping. Anodized aluminum mated with filled teflon is sometimes used as an alternative.

BEARINGS

As mentioned, the lack of piston side forces in free-piston Stirling engines encourages the use of gas bearings to escape the problems of oil contamination of the working space, particularly the regenerator. In gas bearings the working fluid itself is used as the lubricant.

There are two types of gas bearings:

a) hydrostatic,
b) hydrodynamic.

Hydrostatic bearings are externally pressurized. They require a continuous supply of pressurized gas for their operation and the provision of adequate 'drains' to allow escape of the expended gas. Hydrodynamic bearings are self-acting with relative motion between the bearing and the journal generating and sustaining the fluid film separating the sufaces. Hydrostatic bearings are clearly capable of greater loadings than hydrodynamic bearings for the supply pressure of the external fluid can be independently chosen.

Hydrodynamic Bearings

Hydrodynamic bearings have the great advantage of simplicity. They are self-acting and require no external support facilities. They have the disadvantage that the two surfaces are initially in contact when starting and some surface contact dragging is inevitable to establish the hydrodynamic lubricant film. Such occasional surface contact can often be accommodated by the use of compatible hard surface materials, often superficial surface coatings applied by flame spray or plating. Compatible rubbing pairs include, a) chrome oxide and hard chrome plate, b) alumina and alumina, c) carbide on carbide, d) anodized aluminum and hardenable stainless steel.

The linear bearings of reciprocating pistons and displacers clearly present an outstanding challenge to the designers of hydrodynamic bearings. At both ends of the stroke the moving element stops and reverses direction. The hydrodynamic film is destroyed at the moment of reversal and some initial movement must occur to re-establish the film on the return stroke. There is therefore regular and inevitable surface contact at the ends of the stroke.

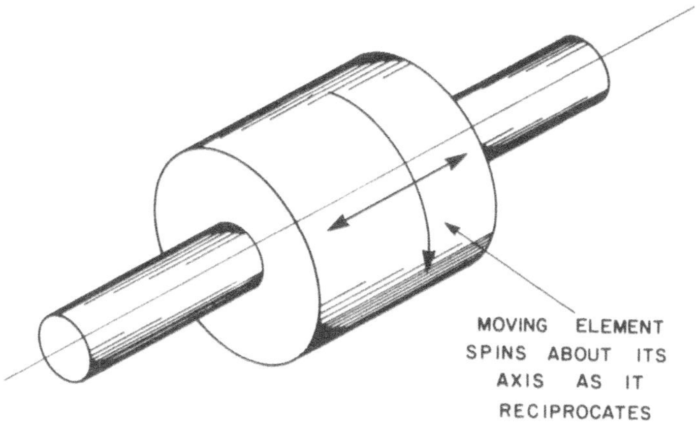

Figure 2.65 The rotary-stroker concept for reciprocating hydrodynamic support.

One solution is to cause the reciprocating elements to steadily rotate about their axis when reciprocating as shown in Figure 2.65. The rotary action generates a hydrodynamic film and maintains separation of the surfaces even at the instant of reversal. This approach was adopted by Breckenridge *et al* (1971) for the 'rotary-stroker' cryocoolers and other machines developed by A.D. Little for the U.S. Air Force. Similar work involving spinning reciprocating elements and an alternative magnetic suspension were discussed by Gasser *et al* (1979) and Gasser *et al* (1981) in connection with long-lived Stirling cryocoolers for space applications.

Hydrostatic Bearings

The elements of a hydrostatic gas lubricated bearing are shown in Figure 2.66. The shaft is supported in a bearing journal by a supply of gaseous lubricant supplied to the bearing at the center of the journal and admitted to the bearing through a ring of admission orifices. Lubricant drains are located at the extremities of the journal and coupled to a return line. Sometimes formal drains are not provided and the fluid simply escapes from the ends of the bearing.

A necessary condition for proper operation of the bearing is that the pressure at inlet is substantially above the exit pressure. The fluid pressure distribution in the journal is idealized in Figure 2.66 to a linear relationship. In practice it is likely to be anywhere between the two broken lines at the right of the figure. The upper broken curve is indicative of a bearing carrying a very high load with minimal clearance between the shaft and the journal or an excessive gas supply through oversize ports. The lower curve is characteristic of a bearing starved of fluid with constricted or

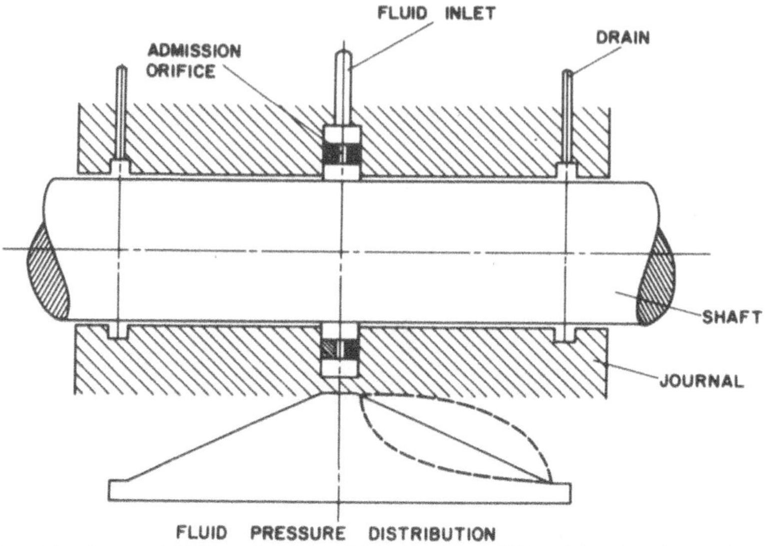

Figure 2.66 Elements of a hydrostatic gas bearing.

blocked admission ports or with excessive clearance between the shaft and the journal; this bearing could sustain only light loads.

Compared with liquids, gases are highly mobile because of their low density and viscosity. This is particularly true for the low molecular weight gases hydrogen and helium, often used as the working fluids in Stirling engines. The reasons for this preference are discussed in detail elsewhere (Walker (1980)) but essentially it is because they have a combination of thermophysical properties that facilitates the transfer of heat with relatively low pumping work. The characteristics that are attractive from the thermodynamic aspect are unfortunately not congruent with the characteristics best suited for gas lubricated bearings. Helium and hydrogen would not be the first choice for the fluid medium in a gas-lubricated bearing all other things being equal.

The use of gas rather than liquid lubricant and, moreover, the customary use of the light gases demands a high order of precision in the manufacture of gas bearings to minimize the gas flow requirements. The clearance of the shaft in the journal will be the smallest achievable at the limits of manufacturing technology, typically less than 0.05 percent of the shaft diameter (5/10000's of an inch per inch diameter) with the mating surfaces customarily finish ground, or ground and lapped. The surfaces are commonly finished as hard as possible to avoid surface damage in brief contacts during the initial start-up or when overloads are suddenly applied. Flame-sprayed carbide, furnace carburizing or hard anodizing are recommended with post-surface treatment finishing to the final fit.

The extremely close clearances necessary in a gas lubricated bearing lead to a very rigid shaft with virtually zero radial movement. This facilitates the use of very close tolerance gas seals in place of mechanical rubbing seals with all their attendant friction and wear problems. A principal advantage of the close tolerance seal is total elimination of the frictional work of a dry rubbing seal. Equally important, is that the wear debris, an inevitable result of dry-rubbing seals, is also eliminated. Seal detritus entering the extremely small apertures of a gas bearing acts as an abrasive with catastrophic results.

A typical use of gas bearings to support the displacer of a centerpost arrangement is shown in Figure 2.67. The centerpost and close clearance bearings provide great control and stability for the displacer with unrestrained vertical reciprocating motion and virtually no radial or rocking motion. Close tolerance gas seal elements are incorporated at the lower end of the displacer. The gas bearings are energized by high pressure gas supplied from a reservoir charged by a reciprocating compressor around the centerpost and the lower end of the

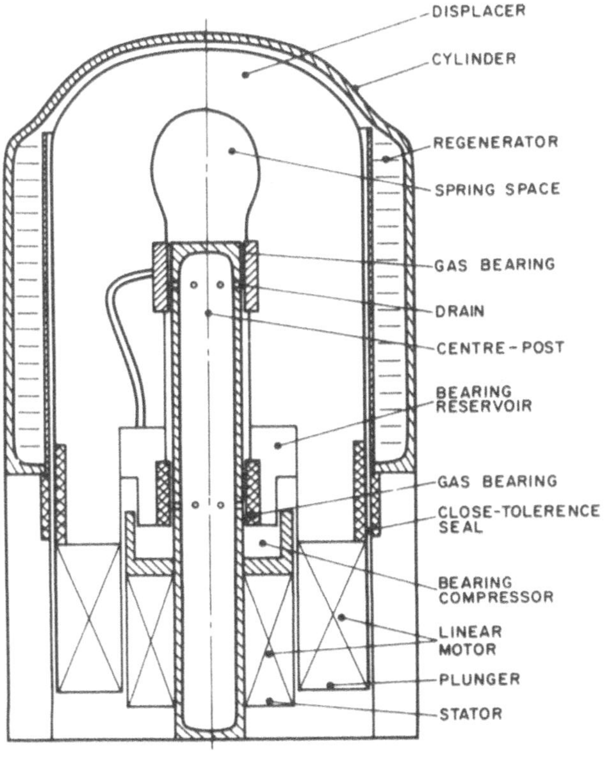

Figure 2.67 Gas bearings supporting the displacer in a centerpost arrangement.

displacer. Another possibility is to bleed high pressure gas from the spring space during maximum compression. The bearings drain to the hollow centerpost and hence to the bearing compressor or simply to the working space below the displacer. The arrangement shown is the hot cylinder of a Vuilleumier cryocooler and includes a linear motor to activate the displacer but of course many other configurations and arrangements are possible.

The design of hydrostatic gas bearings for high performance systems is a specialized activity. The principles are easily understood and applied to simple small engines but designs for high loading are best left to the experts. There are several excellent books on the technology and a wealth of technical reports and papers. Sternlicht (1964) gave an interesting and entertaining introduction to the field that remains worthwhile although now dated. Habercom (1976) (1980a) and (1980b) has given bibliographic data, including abstracts, for 750 papers and reports on gas bearings in the United States Government National Technical Information Service (NTIS) and the Engineering Index (EI) data bases.

MATERIALS

As in most engineering situations the question of materials is a key concern. If properly addressed it would require the remainder of this book and more. We have written at length elsewhere about materials on Stirling engines, (Walker (1980), (1981)) and so will be content to enumerate here just a few of the more important aspects.

The hot parts pose the most demanding materials problems in Stirling engines. If thermal efficiency is important, and required to be high, then the maximum working fluid temperature must be made as high as the materials of the hot parts will withstand. In this regard the Stirling engine shares with the gas turbine and the steam engine the characteristic that the hot parts are exposed *continuously* to the maximum temperatures and stresses. In reciprocating combustion engines the maximum cycle temperatures are achieved only momentarily and can therefore be very much higher than temperatures customarily regarded as the metallurgical limit.

Stainless steels are conventionally used for the general run of heater heads in Stirling engine machines. This is largely because stainless steel is readily available in various forms and because of the comparative ease of fabrication. Type 316 has the best creep properties of the stainless steels commonly available, having about twice the 10000 hour, 650°C rupture strength of, say, Type 304 stainless steel. Of the super-alloys the iron-based Type N-155 has about twice the 10000 hour, 650°C rupture strength of Type 316 stainless. The precipitation-hardened nickel-based alloy Udimet 700 has a 10000 hour 650°C rupture strength about four times that of Type 316 stainless.

This is an overly simple criterion of course for the integrity of a heater head depends very much on the elimination of local hot spots and also the reduction of thermal stresses, occurring perhaps on starting or closing down, that lead to thermal cracking.

Recent developments (Meijer (1978)) have resulted in tubular heaters having manifold headers at both ends of the tubes permitting a single axial penetration of the hot expansion space as shown in Figure 2.68. This is coupled with the use of thermally insulating glass ceramic inserts in the cylinder to permit the use of water cooled expansion space cylinders of relatively low cost steels. The expensive temperature resistant materials are concentrated in local hot regions and used in relatively simple tubular forms.

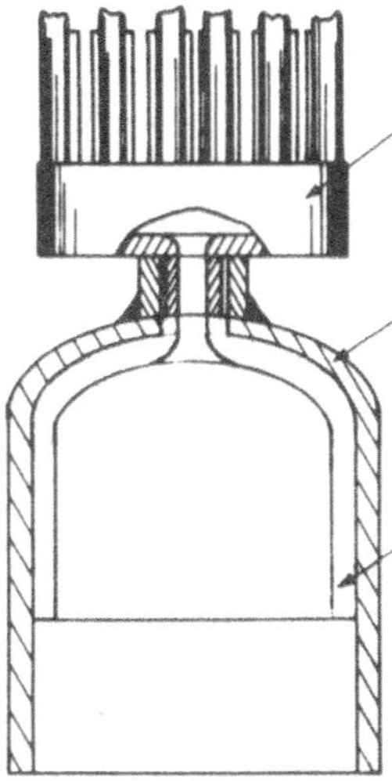

Figure 2.68 Heated head design for single axial penetration of the expansion space cylinder.

Efforts are in progress to incorporate ceramic components in the high temperature regions of Stirling engines but most authorities agree this is a desirable but relatively long-term venture likely to be in general use no sooner than the turn of the century unless there are dramatic and presently unforeseen developments in ceramics technology.

Away from the hot regions materials requirements in Stirling engines are less demanding. The use of aluminum is widely favoured because of its low density, ease of fabrication, relatively low cost, high thermal conductivity and the availability of the hard anodizing process to provide a hard surface finish. Aluminum has a high coefficient of thermal expansion and is, of course, soft and ductile compared with steel so that care must be exercised in its application.

Plastics, particularly filament reinforced epoxy materials appear to have a bright future for the cylindrical shell shapes commonly found in free-piston Stirling engines but so far have not been widely employed.

SPRINGS

Springs are essential in free-piston Stirling engines to provide the restoring forces to maintain cyclic operation of the reciprocating elements. Commonly several springs are incorporated to couple the piston and displacer to the cylinder (ground) or to each other.

Two types of springs are used:

a) mechanical springs,
b) gas springs.

In machines of substantial capacity the spring forces involved are so great that mechanical springs would need to be so massive as to dominate the design. Therefore mechanical springs are not widely used to produce the principal restoring forces but are sometimes incorporated for supplementary purposes to assist reversing, starting or static support.

Mechanical springs have the advantage compared with gas springs of low losses but are subject to failure by metal fatigue and can produce unbalanced side loads on systems ostensibly subject only to axial forces and movements. Gas springs have no side loading and are not subject to fatigue. However, the hysteresis loss in gas springs can be appreciable and there may also be leakage losses. Furthermore, gas springs do not provide positive suspension when the engine is not operating.

Spring Stiffness

The most important characteristic of any spring is the *spring stiffness* K, also known as the *spring constant*. This is the force that must be imposed to cause unit deflection of the spring, i.e. Kg/m or lb_f/in. The spring stiffness has an important effect on the natural frequency of a vibrating system. In a simple spring mass system such as that shown in Figure 2.31 the natural frequency

$$f_n = (1/2\pi)\sqrt{K/M}$$

where f_n = natural frequency,
K = spring stiffness,
M = mass.

For a gas spring such as that shown in Figure 2.69 the spring constant can be reasonably approximated by the equation

$$K = (\gamma p A^2)/V$$

where K = spring constant,
γ = ratio of the specific heats (C_p/C_v) of the spring gas,
p = mean pressure of the spring gas,
A = gas spring piston area,
V = mean volume of the gas spring.

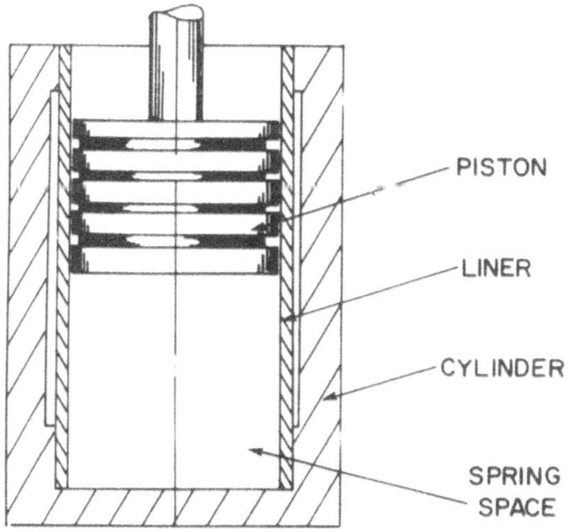

Figure 2.69 Gas spring system.

The validity of this approximate equation is improved when the amplitude of the volume change of the spring (ΔV) is small compared with the total spring volume, i.e. $V/\Delta V > 10$. When the amplitude of the volume change becomes appreciable the gas spring adopts an increasingly non-linear character and the spring force is not a linear function of the displacement. The spring hysteresis losses are also a strong function of the ratio $V/\Delta V$ and when efficiency is a major consideration, provide a compelling reason to maintain a high volume ratio.

CLOSURE

In the chapter we have covered in superficial fashion, much of the special technology associated with free-piston Stirling engines. We have seen how energy flows from high temperature to low temperature and the phenomena of resonance in mechanical systems combine to sustain a free-piston Stirling engine in a steady fixed vibration mode.

There are many possible design variants of free-piston Stirling engines. Some attempt was made to establish a rational basis for classification of the many different arrangements known.

Basic aspects of mechanical vibrations were reviewed and the technique of vector representation was applied to the principal types of free-piston Stirling engines.

Finally, some of the more important practical aspects of free-piston Stirling engines were briefly reviewed. Topics addressed were methods of stabilizing the piston location, seals, bearings, materials, and some aspects of springs.

REFERENCES

Ackermann, R.A. (1981). Dynamic Analysis of a Small Free Piston Resonant Cryorefrigerator. Paper No. 7, pp. 57-69, NBS Special Pub. 607, Refrigeration for Cryogenic Sensors and Electronic Sensors, U.S. National Bureau of Standards, Washington, D.C., May.

Agbi, B. (1971). Beale Free-Piston Stirling Engine. M.Sc. Thesis, University of Calgary, Alberta.

Beale, W. (1976a). Free-Piston Stirling Engines. Notes for UCLA Short Course on Stirling Engines. Univ. of Calif. at Los Angeles, Sept.

Beale, W. (1979). A Free Cylinder Stirling Engine Solar Powered Water Pump. Proc. Int. Solar Energy Soc. Cong., Panta, Ga., June.

Beale, W., Holmes, W., Lewis, S. and Cherg, E. (1973). Free-Piston Stirling Engines - A Progress Report. Soc. of Auto. Eng., Paper No. 730647, Powerplant Mtg., Chicago, Ill., June 18-22.

Benson, G.M. (1980). Analysis and Design of Free-Piston Stirling Engines: Thermodynamics and Dynamics. Proc. 15th Int. Soc. Energy Conv. Eng. Conf., Paper No. 809334.

Breckenridge, R.W., Heuchling, T.P. and Moore, R.W. (1971). Rotary-Reciprocating Cryogenic Refrigeration System Studies. Part I Analysis. Tech. Report AFFDL-TR-71-115, Part I, Wright Patterson Air Force Base, Dayton, Ohio.

de Hartog, J.P. (1956). Mechanical Vibrations. 4th Ed., McGraw-Hill Book Co., New York.

Fokker, H. (1978). The Description of the Stirling Cycle in a Vector Diagram. Proc. 13th Int. Soc. Energy Conv. Eng. Conf., Paper No. 789112.

Gasser, M.G., Sherman, A. and Beale, W. (1979). Developments Toward Achievement of a 3-5 Year Lifetime Stirling Cycle Refrigerator for Space Applications. Proc. Cryo. Eng. Conf., Madison, Wis. (see also Adv. in Cryo. Eng. 1980, Vol. 26 (Ed. K. Timmerhaus), Plenum Press, N.Y.).

Gasser, M.G., Sherman, A. and Beale, W. (1982). Developments Toward Achievement of a 3-5 Year Lifetime Stirling Cycle Refrigerator for Space Applications. Paper No. 11, pp. 103-115, NBS Spec. Pub. 607, Refrigeration for Cryogenic Sensors and Electronic Systems, US National Bureau of Standards, Washington, D.C., May.

Grinnell, S.K. (1956). Flow of a Compressible Fluid in a Thin Passage. Trans. A.S.M.E., pp. 765-771, May, (see Paper No. 55-SA-13).

Habercom, G.E. (1976). Gas Bearings: Volume I 1964-1970. Accession No. NTIS/PS-76/0969. NTIS, Springfield, Va.

Habercom, G.E. (1980a). Gas Bearings: Volume II 1971-1979. Accession No. NTIS/PB-80/2499. NTIS, Springfield, Va.

Habercom, G.W. (1980b). Gas Bearings/Engineering Index 1970-1979. Accession No. PB80-802507. NTIS, Springfield, Va.

Martini, W. (1982). Private Communication. Martini Engineering, 2303 Harris, Richland, Washington, U.S.

Myklestad, N.O. (1963). Fundamentals of Vibration Analysis. McGraw Hill Book Co., New York.

Senft, J.R. (1979). Advances in Stirling Engine Technology. Proc. 14th I.E.C.E.C. Paper No. 799252, pp. 1175-1179.

Sternlicht, B. (1964). Design and Application of Gas Bearings. Paper No. 64-MD-3, A.S.M.E., New York.

Tse, F.S., Morse, I.E. and Hinkle, R.T. (1978). Mechanical Vibrations. Allyn and Bacon, Boston, Mass.

Walker, G. (1980). Stirling Engines. Oxford University Press, Oxford.

Walker, G. (1983). Cryocoolers. International Monograph on Cryogenics. Plenum Press, New York.

Wood, G. (1980). Lecture Notes for Free-Piston Stirling Engine Workshop. Sunpower Inc., Athens, Ohio.

CHAPTER 3 - THEORETICAL ANALYSIS AND DESIGN

INTRODUCTION

Stirling engine analysis is difficult. Inside a real engine a continuum of cyclic fluid and thermal processes take place from one end of the machine to the other. Mathematically modelling this complex interrelated assortment of events in such a way that numerical calculations of performance can actually be carried out and yield meaningful useful results is indeed a challenging problem. Accordingly, just as the mountain entices the climber, this problem has attracted the attention of many very able minds. Although several of the major efforts remain proprietary, the open literature on the subject is already extensive.

The complex process within a Stirling engine is caused and sustained by two simple overt actions, namely the movements of the piston and the displacer (or two pistons). For a kinematic engine, this movement is immutably built into the hardware and the analysis begins with piston and displacer motion specified. From here one goes on to the difficult matter of describing what occurs to the gas and metal inside the engine.

The analysis of a free-piston free-displacer engine is further complicated by the fact that the dynamics of the piston and displacer are caused by the fluid processes their motion causes. This leaves fewer known starting parameters and therefore the analysis is more difficult.

The fact that the dynamics of the free-piston engine are coupled to the fluid thermodynamics makes the designer's task more difficult, and somewhat paradoxically forces greater reliance upon analysis in the design process to achieve a given level of success. Wood (1982)* of Sunpower Inc. puts it this way: "A poorly designed crank engine will usually have less than expected performance, whereas a poorly designed free-piston engine will likely not run at all."

For given piston and displacer strokes and phasing, the same thing happens within the free-piston engine as does inside a kinematic engine. Therefore the many and varied analysis schemes developed for crank engines can be used to model the workspace processes in a free-piston engine as well; one only need add means to determine the piston and displacer dynamics.

This chapter will attempt to provide simply a *guide* to the

---

*Personal communication, 1982.

field of Stirling engine analysis. Following a self-contained presentation of elementary approaches, we shall content ourselves with brief descriptions of the major categories of analysis methods and provide citations of the more accessible books and papers which appear to offer efficient entry points to the literature; these references in turn will open up the remaining body of published works if there is a need to go further. Secondly, the chapter will attempt to point out approaches to problems unique to the free-piston engine, namely predicting stroke, frequency and phasing. This is especially important at the lower levels of analysis since the higher analyses contain enough detail to numerically integrate pressure forces to obtain piston and displacer dynamics.

ELEMENTARY DESIGN TOOLS

Ideal Cycle

In the early stages of an engine design project, one wishes to simply and easily predict power and efficiency from gross engine dimensions and operating conditions. The obvious starting point is from the ideal Stirling cycle shown in Figure 1.1 in the p-V plane. One easily calculates the cyclic work:

$$W = \oint pdV = mR\ (T_E - T_C)\ \ln\ (V_1/V_2) \tag{3.1}$$

This then could be used with estimated speed to predict engine power:

$$P = fW$$

where f is the cyclic frequency.

This scheme, however, has two major shortcomings. First, formula (3.1) is valid for the ideal Stirling cycle, and any practical machine is quite far from the ideal for the reasons discussed in detail in Chapter 1, e.g. non-isothermal processes, continuous piston motion, deadspace, etc. The practical engine cycle is at best a severely rounded version of the four-cornered ideal pV diagram.

One could attempt to account for these departures from ideality and for mechanical friction losses by taking an appropriate fraction of W. Since this multiplier can only be surmised from practical experience, Martini (1978) discreetly refers to it as an "experience factor". This is quite workable in practice actually if one does not expect a grand "universal factor" applicable to all manner of engines under all conditions. If one deals with a group of engines limited in size, type, configuration, and operating conditions, then an experience factor can usually be associated with the engines in this category with acceptable results. An appreciable change in any one of the major features, however, will usually require that a new factor be found. And one always must be willing to allow for an ample margin of error with this approach. But before discussing this further, let us consider the second shortcoming of formula (3.1).

Formula (3.1) does not contain the most *convenient* variables for rapid application and quick comparisons. For example, one nearly always has the bore and stroke of a given engine at hand, but rarely the maximum and minimum volumes, V_1 and V_2. Likewise, one can easily read off the mean cycle pressure of an engine from a gauge on the buffer space chamber, but the mass m of the working fluid is much more elusive.

Beale Number

Apparently motivated by these considerations, William Beale evaluated the following dimensionless quantity B_n for various known Stirling engines:

$$B_n = \frac{P}{fV_o p_m}$$

where
- P = engine net or brake power,
- F = frequency,
- p_m = mean cycle pressure,
- V_o = total volume variation (piston swept volume in the case of a piston/displacer machine).

He found that for successful well designed engines, B_n was very nearly always about .15 in systematic (e.g. SI) units. If one measures V_o in the more convenient units of cc and p_m in bar, then with P in watts and f in Hz, B_n becomes .015.

This discovery quickly became and remains a popular tool for preliminary estimates of engine performance:

$$P = .15\ f\ V_o\ p_m \qquad (3.2)$$

As an example of the ease with which this formula can be used (exactly the reason for its popularity) consider a small air engine of 12 cm bore and 6 cm stroke running at say 1200 rpm with a moderate mean pressure of say 4 bar. Applying (3.2) we find an expected power output in the neighborhood of:

$$P = .015\ \frac{1200}{60}\ (12)^2\ \frac{\pi}{4}\ (6)\ 4$$

$$= 814 \text{ watts}$$

Most of the engines considered by Beale operated more or less between the same temperature extremes of about 650°C for the heater and 65°C for the cooler. Walker (1979) produced a very useful graphical relationship of the Beale number temperature variation shown in Figure 1.14. The central curve is the mean value of the number for all engines surveyed and the outer dashed curves show the range of variation likely to be encountered. Carefully designed engines, particularly with regard to heat exchangers, will fall nearer the upper curve while the lower curve is characteristic of the simpler engines.

West (1981) discovered the analog of formula (3.2) taking this temperature effect into account. He found:

$$P = F\ f\ V_o\ p_m \frac{T_E - T_C}{T_E + T_C} \tag{3.3}$$

for typical real Stirling engines where T_E is the hot end temperature and T_C the cold. The value of the experience factor F suggested by West is .35 (.035 for the convenient mixed units) following Martini's observations. This value produces a Beale number of .15 (.015) for a 650°C by 90°C engine which agrees well with the original formula (3.2).

Senft (1982) subsequently deduced the form of (3.3) directly from the ideal Stirling cycle with no additional assumptions or approximations. This proves that the formula is intrinsically applicable to all Stirling engines. For the ideal cycle F = 2.

Formula (3.3) can thus be used in lieu of (3.2) to cover Stirling engines operating at very high heater temperatures (e.g. ceramic hot end) and at low temperatures (e.g. Fluidyne). For example, if we were to operate the engine in the above example at the temperatures T_C = 90°C and T_E = 550°C, the output would drop to:

$$P = .035\ \frac{1200}{60}\ (12)^2\ \frac{\pi}{4}\ (6)\ 4\ \frac{823 - 363}{823 + 363}$$

$$= 737 \text{ watts}$$

about a 10% power reduction. Of course, in practice, the effect would probably be greater due as well to a decrease in speed with the lower heater temperature.

Although it should be clear from the preceeding discussion and examples, it is most convenient to use the heater and cooler metal temperatures for T_E and T_C in formula (3.3). At this level of analysis one usually cannot do anything but guess at the *gas* temperatures achieved; it is best to let the experience factor account for these differences (which in a well designed engine are rather small) and accept the results as a first (but usually surprisingly good) guide.

The remarks made earlier regarding the restricted range of validity of an "experience factor" apply to formula (3.3). After all, for the ideal cycle the power given by (3.3) is identical to that calculated via formula (3.1) (Senft (1982a)). Thus in particular if one wishes to work on low T_E engines, the validity of F = .35 should be checked and quite probably will have to be modified.

Frequency Estimation

The most uncertain value involved in applying the mean pressure power formula (3.3) is the frequency f. This is particularly true for kinematic engines where at this level of analysis one must for the most part rely on intuitive judgements based upon experience with real engines. Especially on new designs, a good maxim to follow is "be safe - guess lower". As can be readily appreciated, a much higher level (order) of analysis incorporating at least internal losses and external loading is required to accurately predict engine frequency.

Since the free-piston free-displacer Stirling engine is nearly a constant frequency device, it is much more amenable to estimation of its frequency than is the fully kinematic engine. For a first approximation of the frequency one may take the natural frequency of the piston mass and spring combination:

$$f = f_n = \frac{1}{2\pi}\sqrt{K/M} \tag{3.4}$$

where M = total piston mass,
K = net spring constant on piston.

It should be noted that M is the total mass of the piston assembly, including any part of the load device that is attached to and reciprocates with the piston. K is the sum of the constants of all springs coupled to the piston assembly. This includes the "spring effect" of the workspace. The constant of the workspace gas spring can be estimated by the following formula for the constant K_g of a gas spring:

$$K_g = \frac{\gamma p_m A^2}{V_m} \tag{3.5}$$

where $\gamma = C_p/C_v$,
p_m = mean gas pressure,
A = piston cross-sectional area,
V_m = mean volume of spring chamber.

Of course an adiabatic gas spring does not have a "constant", but for a gas spring with a small volume variation relative to the mean volume, Hooke's law with the above calculated K_g is essentially valid.

Much more goes on within the workspace of a free-piston Stirling engine than does inside an adiabatic gas spring, so that (3.4) is usually only a starting point, but a valuable one nonetheless. If the engine is fitted with a large additional spring on the piston, then the workspace spring constant is less critical. As a rough rule of thumb for typical engine configurations, Beale suggests reducing the value given by (3.4) by 20% (Beale (1976)).

One should keep in mind that although the free-piston engine runs essentially at a fixed frequency the piston stroke is not fixed. The swept volume V_o in fact varies with load and mean pressure. Thus formula (3.3) offers no more for the crank or the free engine; in the first frequency is indeterminate (at this elementary level) and in the second swept volume is indeterminate.

An interesting consequence of the discussion above is that the mean cycle or charge pressure can be used within limits to modify engine operating characteristics. For example, in an engine with very light external piston springing, the workspace gas spring effect prevails. Thus by equations (3.4) and (3.5), the engine operating frequency is proportional to the root of the mean pressure:

$$f \sim \sqrt{p_m} \tag{3.6}$$

Thus one can manipulate frequency by varying the charge pressure, which may be required to match the optimum operating frequency of the load device.

Charge pressure can also be used to vary the piston stroke to some extent. This is most effective when the machine operates near the power output limit imposed by its external (and internal) heat exchangers. If mean pressure is increased, frequency will increase as equation (3.6) predicts and if power cannot increase proportionally because of heat flow rate limits, then the swept volume must decrease.

Efficiency

As was mentioned in Chapter 1, the efficiency of the ideal Stirling cycle *with regeneration* is:

$$\bar{\eta} = 1 - \frac{T_C}{T_E} \tag{3.7}$$

This is the well known Carnot efficiency which is the maximum possible efficiency for any engine operating between the temperatures T_E and T_C. This is because the amount of heat *absorbed* by the working gas during one constant volume process is exactly the same as the amount *rejected* in the other constant volume process. This makes regeneration - the alternate temporary storage and release of this heat - theoretically possible, and in practice quite effective regenerators can be made.

So although (3.7) is theoretically appropriate for Stirlings, in practice it is much too optimistic for any use beyond that as an upper bound (which it is for *all* engines). In addition, in practice one is usually more interested in *brake efficiency* which is the ratio of the power available at the shaft to the rate of heat absorbed by the engine. Specifically this latter quantity is defined as the sum of the rate at which heat is absorbed by the engine cooling water and the measured shaft power. This definition is used because it is more convenient to measure the heat delivered to the cooling water than it is to measure (somehow) the heat delivered to the hot end of the engine, and convective heat loss from hot engine parts to the surroundings can be easily minimized by insulation, at least for the purpose of making this measurement.

A realistic approach to estimating brake efficiency is to parallel the one employed above, namely to multiply $\bar{\eta}$ by an experience factor, derived from data on similar engines. Martini (1978) surveyed data from many modern engines (Philips, United Stirling, General Motors, MAN/MWN) and found that brake efficiency ranged from 28% to 69% of $\bar{\eta}$, the Carnot efficiency; the values of $\bar{\eta}$ were calculated from the heater and cooler *metal* temperatures. Thus a value of about .5 of Carnot efficiency could be expected on the average for a modern well designed engine.

Little more can be said at this level to improve the accuracy or certainty of the power and efficiency calculation methods described above. To do this one must advance to an *analysis scheme* that models the engine "globally", that is that takes into account the important interactions between the various components of the engine.

FIRST ORDER ANALYSIS METHODS

It is customary in the field to categorize analysis schemes into three groups or "orders". In practical terms, the order is simply determined by the complexity of the analysis which is largely a function of its fidelity or, equivalently, which is inversely related to the number of idealizing assumptions employed. The third order schemes make the fewest idealizations and therefore in principle produce the highest level of prediction accuracy. At the other end of the spectrum is the first order analysis.

Schmidt Analysis

Although many first order analysis schemes have been put forth in the literature, they invariably possess a common core of assumptions first proposed by G. Schmidt (1871):

i) the working fluid is an ideal gas,
ii) the mass of gas in the engine workspace is constant,
iii) the instantaneous gas pressure is constant throughout the workspace,
iv) the workspace consists of isothermal regions,
v) the piston and displacer move sinusoidally.

Assumption i) is quite reasonable and ii) can be closely approximated in practice by carefully fitted piston and displacer rod seals (and of course sound static seals as well). Assumption iii) ignores fluid friction pressure drop which was probably quite acceptable in Schmidt's day of slow running engines. This pressure drop is significant for a high speed engine, however, so the assumption limits the power prediction accuracy of the analysis in that case.

By far the most significant assumption of the lot is iv), so much so that this analysis is also often referred to as an *isothermal analysis*. The workspace is usually taken to consist of three regions: the expansion space, the compression space, and a void or dead space. The gas in each of these spaces is assumed to be at a constant temperature associated with that space: all the gas in the expansion space is at the fixed temperature T_E, all gas in the compression space at a lower temperature T_C, and gas in the dead space is at some intermediate temperature T_D (usually the average of T_E and T_C).

This assumption implies then that the compression and expansion processes are isothermal, which is in perfect accord with the ideal cycle but which is quite far from the practical machine in which the processes are nearer to adiabatic. The assumption also requires that gas displaced from one space to another instantly changes temperature to match the new space. This of course is physically impossible, requiring infinite heat transfer rates or zero specific heat.

However, accepting this assumption produces a true global model that is mathematically tractable, which makes it of great value as a basic guide to Stirling engine design.

These first four assumptions completely determine the workspace pressure as a function of piston and displacer position. Figure 3.1 shows the basic Stirling engine capsule and illustrates the terminology used with our isothermal assumptions; the nomenclature table further describes the terms. The workspace pressure is:

$$p = \frac{mR}{\frac{V_c}{T_C} + \frac{V_e}{T_E} + \frac{V_D}{T_D}} \tag{3.8}$$

Note that the only variables on the right hand side of (3.8) are V_e and V_c, which are linear functions of displacer and piston position.

If the piston and displacer are made to move cyclically, then V_e and V_c will be periodic functions of crank position or time and the work per cycle is:

$$W = mR \oint \frac{d(V_e + V_c)}{\frac{V_c}{T_C} + \frac{V_e}{T_E} + \frac{V_D}{T_D}} \tag{3.9}$$

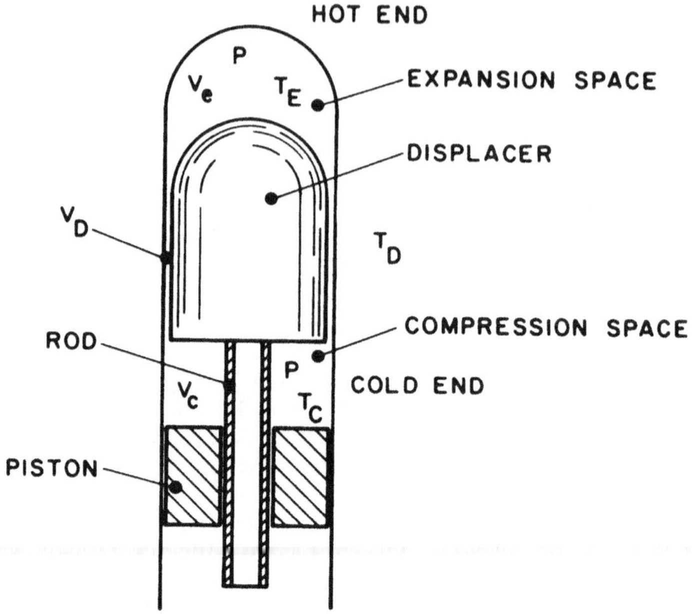

Figure 3.1 The basic Stirling engine capsule modelled isothermally.

where the integration is carried out over the common period of V_e and V_c. Note that (3.9) is completely general; it applies for all manner of piston and displacer motions. W may be easily determined numerically for any given periodic V_c and V_e functions.

Assumption v), that the piston and displacer move sinusoidally with a common period and a fixed phase angle separation, is usually acceptably close to the real engine. In the kinematic engine relatively long connecting rods and massive flywheels produce near harmonic motion, and in a free-piston free-displacer engine running without collisions, and against a damper type load, the motion is also close to harmonic after steady state is achieved. Common exceptions occur in the Ringbom engine running in the *over-driven mode* (Chapter 5) and in kinematic engines with novel drives such as the rhombic (Senft (1982b)).

Schmidt (1871) showed that with assumption v), it is possible to actually integrate (3.9) and thus obtain a closed-form expression for the cyclic work W in terms of the basic engine geometry and operating parameters. Since then many have contributed to the literature refining and extending Schmidt's basic results to cover the full range of Stirling engine configurations from various points of view. Rather than attempt to survey or repeat here what is readily available in many forms from many sources, we refer the reader to Martini (1978) and Walker (1980) for more detail and guidance to the literature.

The Schmidt cyclic work formulas have proved to be powerful tools for gaining insight into the basic design problems of the Stirling. Their principal utility in this direction has been to find optimum values of phase angle and swept volume ratio with respect to maximizing specific power. This work followed long after Schmidt's work, with a resurgence of serious interest in the Stirling and depended upon the electronic computer to carry out the optimization of the otherwise impenetrable work function. The pioneering efforts in this are due to Finkelstein (1960), Walker (1962) and Kirkley (1962).

Free-Piston Engine Dynamics

Although the free-piston engine running in non-collision steady state meets the five assumptions above as well as does the kinematic engine, piston and displacer strokes and phase angle are not known from the blueprints for a free-piston engine as they are for the kinematic. This makes applying the Schmidt results impossible without further analysis. Complicating the matter is the fact that not only are these quantities unknown, but they are variable with load and intricately coupled. Thus one must begin by determining to the extent possible the *dynamics* of the piston and displacer given the hardware description, temperatures, load, damping, springing, etc. One is thus led to a system of second order differential equations.

To illustrate what one can expect, consider the equations for an engine having negligible canister motion as schematically represented in Figure 3.2. Assuming the engine can be adequately modelled with linear springs and velocity proportional dampers, the equations of motion for the piston and displacer are:

$$M_P\ddot{x}_p + (C_P+C_R)\dot{x}_p - C_R\dot{x}_d + (K_P+K_R)x_p - K_Rx_d = (p-p_o)A_P$$

$$M_D\ddot{x}_d + (C_D+C_R)\dot{x}_d - C_R\dot{x}_p - K_Rx_p + (K_D+K_R)x_d = (p-p_o)A_R \quad (3.10)$$

with all the coefficients on the left sides constant.

Unfortunately, the form of the pressure function given by equation (3.8) precludes a closed form solution of the system (3.10). A computer based numerical solution of (3.10) is of course in principle a simple matter. However, careful thought must be given to how displacer/piston and displacer/cylinder collisions will be handled. These are virtually certain to occur the first several times an engine is simulated and in fact will often stubbornly persist when radical designs are attempted. Since one is primarily interested in steady state operation without collisions, one approach is to program to modify initial conditions when a collision occurs and restart. Another strategy relies primarily upon updating displacer velocity when contact occurs with the piston or the canister to match the velocity of the latter; since the displacer is made as light as possible, this is a reasonable treatment. Masses and springs are adjusted until the

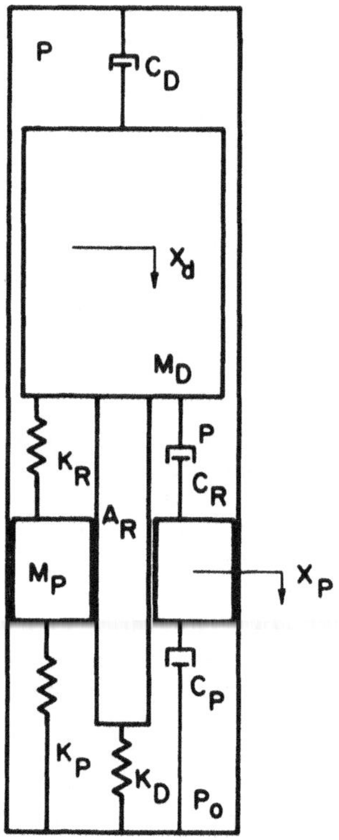

Figure 3.2 Schematic representation of a free-piston engine.

desired operation occurs. The results can then be used in the Schmidt work formula discussed above.

In any event, the greatest value of such a program is as a quick guide to getting the dynamics of a proposed engine within desired limits. It is desirable therefore that the program be a fast running interactive one. An analog computer is well suited to this task in fact, and is relatively inexpensive; L. Mitchell designed and constructed a very effective version of one of these devices at Sunpower, Inc. in 1979.

Linearization of Pressure

Taking the buffer space pressure p_o equal to the workspace pressure p given by (3.8) at the midstrokes of the piston and displacer, the right hand sides of the equations (3.10) can be easily approximated by a linear function of x_p and x_d, namely by the total differential of p considered as a function of x_p and x_d. Thus:

$$p-p_o = S_p x_p + S_d x_d$$

where

$$S_p = \left.\frac{\partial p}{\partial x_p}\right|_{(0,0)}$$

and (3.11)

$$S_d = \left.\frac{\partial p}{\partial x_d}\right|_{(0,0)}$$

This approximation was first suggested by Berchowitz and Wyatt-Mair (1979).

This makes the linear second order system (3.10) homogeneous with constant coefficients and thus highly manageable. Let us simplify notation by representing this system by:

$$\ddot{x} = A\dot{x} + B\dot{y} + Cx + Dy$$
$$\ddot{y} = a\dot{x} + b\dot{y} + cx + dy \tag{3.12}$$

where A, B, C, D, a, b, c and d are constants, $x = x_p$ and $y = x_d$.

The form of (3.12) suggests simple sinusoidal functions for a steady state solution, which of course is our principle interest. Therefore, we make the following substitutions into (3.12):

$$x = X \sin \omega t$$

$$y = Y \sin (\omega t + \phi)$$

where X and Y are the amplitudes, ω is the angular frequency, and ϕ is the phase lead of the displacer over the piston.

It is an elementary, though lengthy matter, to obtain closed form solutions for the frequency and phase angle. We find:

$$\omega = \sqrt{\frac{aD+cB-Ad-Cb}{A+b}} \tag{3.13}$$

and

$$\tan \phi = \frac{aB-Ab}{A\omega^2+Ad-aD}\,\omega \tag{3.14}$$

The amplitudes X and Y are not uniquely determined in this analysis scheme, but are strongly related through the frequency and phase angle by equations resulting from the substitution. For example, we find:

$$A\omega X + B\omega Y \cos\phi + DY\sin\phi = 0 \qquad (3.15)$$

which can be used to find X once Y is surmised; one can always adopt the optimistic viewpoint here and take Y to be half the desired displacer stroke! In any event, equations (3.13), (3.14) and (3.15) resulting from this linearized analysis constitute a quickly applied design tool for free-piston engines requiring only a hand held calculator. After solving for frequency, the phase angle and piston and displacer strokes can be determined. At this point the standard Schmidt equations (Martini (1978)) can be used to find cyclic work. This with the frequency yields the engine power. Of course, first order power predictions are highly optimistic because of the simplifying assumptions taken, and must be considerably reduced if a realistic estimate of actual engine output is desired. Martini (1978) has suggested an "experience factor" of 0.3 to 0.4 for obtaining the brake power of typical Stirling engines from the Schmidt predictions. Wood (1980a) has modified this approach to compensate for the nonlinearities of the engine pressure wave and flow damping to further improve accuracy.

Despite being rather rough for power prediction, the first order analysis above can be invaluable for finding the frequency and phase angle of a free-piston engine. More generally, this analysis also provides some insight to the frequency and phase angle *characteristics* of free-piston engines. For example, Berchowitz and Wyatt-Mair (1979) have investigated the dependence of frequency on load in two special cases with interesting results. There appears to be scope for much more work in this direction, sure to lead to results useful for guiding the designer and aiding our understanding of the behavior of free-piston Stirling engines.

SECOND ORDER ANALYSIS

Analysis schemes beyond the first order level result when one diminishes the effects of the Schmidt assumptions discussed above. This can be accomplished in essentially three ways. First, one can remove one or more of the principle assumptions. Of course, assumption i) would be retained, and v) would apply reasonably well to most free-piston engines (except at start-up or under unusual running conditions). Assumption ii) is the least significant of the remaining, so one would naturally look to dropping iii) and iv) first. Although this is the most obvious way to improve the fidelity of the model, it more or less instantly produces a very complex and computationally difficult model. In fact in practical terms, one is essentially at the third order level, at which point it is not noticeably more difficult to drop v) and ii) as well!

The second approach to an improved model is to modify one or more of the assumptions to bring them closer to what one suspects actually goes on in the engine. For example, adiabatic expansion and compression spaces with isothermal heater and cooler spaces could well replace the assumption of all isothermal spaces. As with this particular example, this approach appears to generally swiftly remove one from the realm of closed form expressions for the power output, but numerical computation is not nearly as complex or lengthy as in the first approach.

The third approach involves introducing corrections for the idealizations and inadequacies of the assumptions. For example, one can estimate the power lost due to aerodynamic friction by known correlations and reduce the power output already calculated by an appropriate amount; this presumably "corrects" for assumption iii) and improves the accuracy of the final result. This approach when used alone often leaves the basic Schmidt mathematics untouched, and usually employs closed form formulas for estimates of the corrections. Thus an application generally could be carried out on a hand-held calculator if time is not a factor.

The second and third approaches, used separately or in combination, result in what have come to be generally known as second order analyses and many are described in the literature. It would be pointless to attempt to survey all the second order schemes proposed over the years, and there is really no need to do so here. The reader who has advanced in analytical sophistication to the point where a second order analysis is desired, is probably well prepared to consult the literature alone and go on from there to refine, adapt, and possibly improve what has been done to suit the need at hand. As an initial guide to the literature and as a base for general understanding, we shall therefore limit our discussion of second order schemes to a description of two important approaches to basic power calculation and the various corrections required for finding realistic values for actual power output and efficiency.

Basic Power Calculation

As already indicated, assumption iv) is the most idealistic of the Schmidt analysis, especially regarding the compression and expansion spaces. These spaces are usually simple cylinders and therefore heat transfer from the bulk of the gas to the walls is relatively quite limited at typical engine speeds. Therefore one might argue that modelling these spaces as adiabatic would be far closer to reality. Thus Finkelstein (1960a) developed an analysis scheme to model these spaces as adiabatic. In fact, Finkelstein's scheme is quite general and allows one's choice of process in the expansion and compression spaces to be between isothermal and adiabatic as well. The analysis retains some isothermal spaces; heater, cooler and regenerator spaces are assumed isothermal (the regenerator space temperature used is the mean). Although this is still an idealization of the practical machine, it is quite acceptable when coupled with the adiabatic spaces, for now we have a model reflecting the high heat transfer rates in the heat exchangers and the low heat transfer rates in the compression and expansion spaces.

Of course a closed form expression for the cyclic work is impossible at this point, and one must resort to numerical analysis methods via a computer; a ready-to-run program ALPHA WEST by C. West is now available from Westware Co. and through Martini Engineering. The computation is quite fast compared to a typical third order scheme and can be easily done on a personal computer (or even a programmable calculator if time is not of the essence). It is interesting to note that West (1980) has found a closed form solution in the special case where the expansion space is adiabatic and the compression space is isothermal; this is applicable to fluidynes in which the cold space can be "isothermalized".

Walker and Khan (1965) carried out a numerical study of the Finkelstein adiabatic analysis applied to hypothetical engines. Effects of temperature, swept volume, phase angle, and dead volume on engine output and efficiency are quite interesting and appear much more in line with the real engine than do simple first order results. The adiabatic analysis also produces a more realistic picture of efficiency than the Carnot efficiency given at the first order level. This then appears to be an excellent basis for a fast running analysis method which, when suitable corrections are also included, should prove to be a practical tool for the design of real Stirling engines; these corrections will be briefly discussed below. Lee (1981) has reported just such an analysis program with a single case comparison against a sound third order analysis (Sunpower's); the predicted power and efficiency were in quite reasonable agreement for the engine simulated.

Martini (1978) presented an easily applied second order analysis scheme which is based upon the Schmidt calculation of the basic power. Martini introduced a clever reiterative calculation of effective gas temperatures to offset the optimism of the isothermal assumption, and set forth a very clear and thorough method of calculating a realistic efficiency estimate. The work is of great significance and is required reading for anyone interested in second order analysis. It is also an excellent encyclopedia of first order analysis methods, formulas and references.

The strategy employed by Martini's second order scheme starts with the calculation of the basic engine power. This is simply an application of the appropriate Schmidt formula if the piston/displacer motion is sinusoidal. If not sinusoidal, as for example in the rhombic case, one must use numerical integration to evaluate the cyclic work integral (3.9). In this initial computation one may use the heater and cooler metal temperatures as a first approximation to the gas temperatures T_E and T_C. The effective regenerator temperature T_D is the log mean of T_E and T_C:

$$T_D = \frac{T_E - T_C}{\ln T_E - \ln T_C}$$

At this level of analysis one usually wishes to model additional dead space in the hot space and cold space of the engine; this merely entails adding the appropriate constant terms in the denominator of (3.8) and (3.9).

Having gotten the *basic power*, this is divided by the Carnot efficiency $(T_E - T_C)/T_E$ to yield the *basic heat input*. Of course much more energy must be picked up by the gas in the heater to make up for the many losses that occur. Examples of these losses include shuttle conduction, gas conduction, radiation, and reheat loss; these and others will be discussed in greater detail below. This energy loss, is added to the basic heat input (with adjustments for windage power) to yield the *net heat input*, that is the total energy flow that must be delivered by the engine heater to the working fluid. The heat transfer coefficient for the heater type (tubular or annular) is computed, and then the gas-to-metal temperature difference required to effect the net heat input found above is determined. These calculations are based on standard correlations for heat transfer and on an approximate flow time of two-thirds of the cycle time.

A similar calculation is used to find the necessary temperature difference in the cooler. The required heat flow out through the cooler is taken by Martini to be the net heat input less the *net power output*. Net power output is the basic power less windage power and mechanical friction power. Thus the cooler is assumed to remove all the friction energy as would be the case for example with an engine within an insulated enclosure.

With these calculations one has more realistic values for the actual gas temperatures rather than the heater and cooler metal temperatures. Basic power and basic heat input can now be recalculated with these temperatures and then the above can be repeated to yield second (and hopefully better) values for the hot and cold gas temperatures. This is not really difficult since the heat losses already calculated remain the same having been found from the heater and cooler *metal* temperatures. This process can be repeated until the gas temperatures remain essentially the same - only a few times according to Martini for typical cases.

This leaves one with a net power and a net heat input much more realistic than from the Schmidt (first order) analysis. One also gets a good sound estimate of efficiency and a detailed accounting of the various losses within the object engine, which can be of great value in making modifications or design changes. Furthermore, a finished but yet-to-be published second edition of Martini (1978) contains an adiabatic version of the above analysis for improved accuracy.

Loss Mechanisms

Perhaps the greatest contribution to second order analysis of Stirling engines made by Martini's Stirling Engine Design Manual was the very clear and thorough presentation of methods of calculating estimates for the major heat and power losses in a Stirling engine. We shall briefly describe the principal loss mechanisms in these two categories in the interest of providing a general background with particular emphasis on the power losses experienced by the free-piston engine. The reader interested in more detail can consult Martini and the recent literature.

Heat Losses: Conduction

All materials conduct heat to a greater or lesser extent depending on the *thermal conductivity*, a characteristic property of the material. Thermal conduction in Stirling engines is important principally because of heat leakage by conduction from the hot parts along the cylinder and displacer walls. The cylinder and displacer walls are made long and as thin as possible to inhibit conduction.

Shuttle Heat Transfer

Direct thermal conduction effects are enhanced by *shuttle heat transfer* or *the bucket brigade loss*. This is illustrated in Figure 3.3. The effect arises because of the reciprocating action of the displacer. There is a temperature gradient along the length of the displacer and cylinder walls. It may be that the temperatures of adjacent parts of the displacer and cylinder wall are similar when the displacer is at the top of its stroke as shown in Figure 3.4(a). When the displacer moves to the lower position, diagram (b), the equality of temperatures of adjacent parts will not prevail. The temperatures of the displacer will be higher than the adjacent cylinder. Additional heat will be lost by radiation and convective heat transfer because of the temperature difference. The effect is virtually an enhancement of the conduction heat transfer. Rios (1971) contains a detailed analysis.

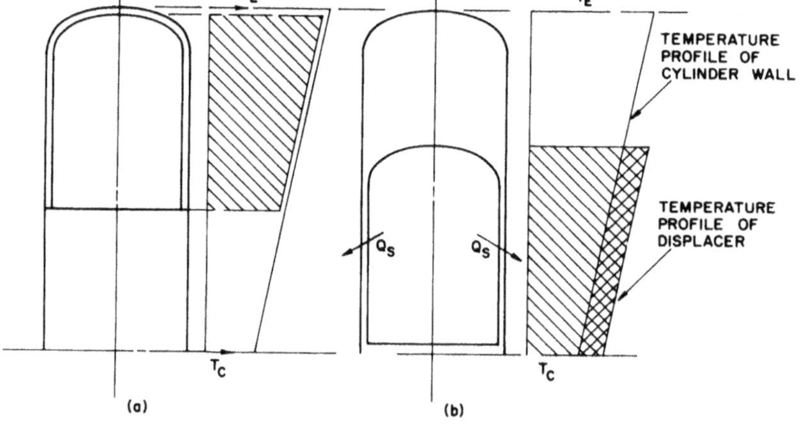

Figure 3.3 Shuttle heat transfer.

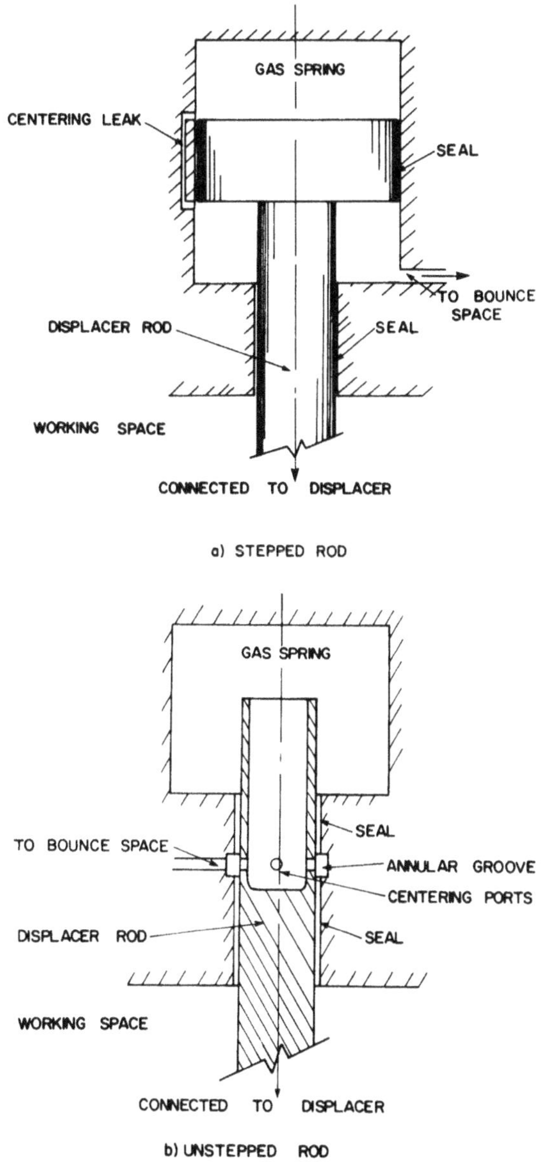

Figure 3.4 Two arrangements of gas springs in free-piston Stirling engines, (after Wood (1980)).

Displacer Annulus Pumping Loss

A similar effective enhancement of the thermal conduction loss occurs as a result of the cyclic filling and draining of the annular space between the displacer and cylinder above the displacer seal, always located at the cold end of the displacer. This cyclic filling and draining occurs as a result of the cycle pressure change.

Radiation and Convection Losses

Various radiation and convective heat transfer losses occur whenever opportunities arise and must be inhibited by the use of radiation shields and enclosures or convective spoilers.

Regenerator Losses

The regenerative heat exchanger is subject to various thermofluid losses. There is an unavoidable fluid friction power loss in the matrix. There is some matrix heat conduction along the temperature gradient. There is some heat conduction in the radial direction to the walls. The heat flux returned to the gas stream in reverse flow is always less than the heat flux to the matrix in the hot blow. There is, therefore, a net enthalpy flux in the matrix from the hot side to cold side. When the heat capacity of the gas per blow becomes significant compared with the matrix heat capacity the regenerator is said to be approaching thermal saturation and regenerative thermal action is impaired.

Power Losses: Flow Loss

Aerodynamic flow losses in the ducts connecting the expansion and compression spaces contribute significantly to the degradation of performance. Pressure losses occur in the heat exchanger tubes and the regenerator matrix. Aerodynamic flow losses are a function of the square of the fluid velocity and may therefore be reduced by generous proportioning of the flow conduits. Unfortunately, excessive dead space is not permissible for the volume and pressure compression ratios are both thereby diminished with consequent reduction in power output.

Seal Loss

The power loss associated with piston seals is due to two causes: friction and leakage. In a free-piston engine, close tolerance seals appear to be the favored seal as discussed in Chapter 2 above, so that friction power loss here is virtually non-existent. Leakage, however, can result in significant power loss as given by the following equation:

$$\text{Leakage power loss} = \frac{\pi\, h^3\, D}{24\, \mu\, L}\,(P_{amp})^2$$

where
- h = radial clearance of piston in cylinder,
- D = cylinder diameter,
- μ = gas viscosity,
- L = length of piston,
- P_{amp} = amplitude of the pressure wave.

This equation refers to a concentric annular leakage path. However, there is a natural tendency for the system to move into a non-concentric position resulting in the crescent-shaped flow path illustrated

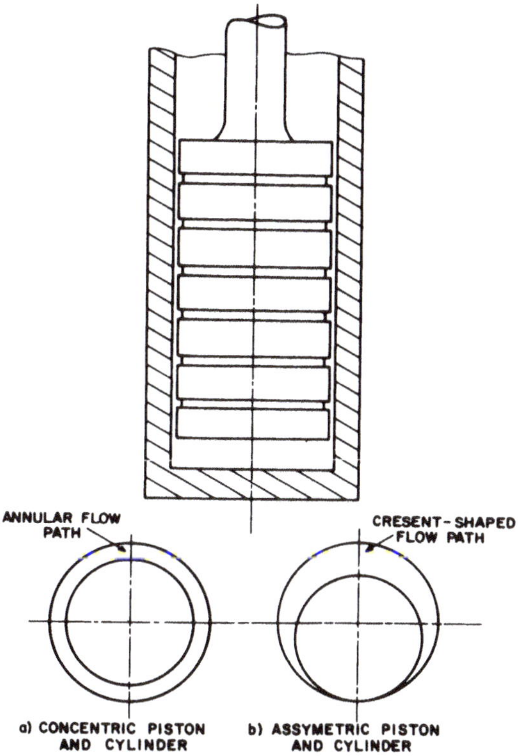

Figure 3.5 Concentric and non-concentric fluid leakage paths.

in Figure 3.5. This results in increased leakage flow and power loss.

Hysteresis Loss

The gas spring, as explained in Chapter 2, is essential to the high performance of a tuned free-piston engine. It is also the source of power loss, the so-called hysteresis loss. It is due to essentially adiabatic gas temperature fluctuation in the interior region of the spring space and to essentially isothermal conditions at the walls. Taking only conduction into account, Breckenridge *et al* (1971) found that the ratio of gas spring power loss to energy stored could be expressed as:

$$\frac{GSPL}{ES} = \frac{F}{2\sqrt{2}} \frac{(\gamma-1)}{V} S \sqrt{\alpha\omega}$$

where GSPL = gas spring power loss,
ES = energy stored,
F = empirical factor determined experimentally,

γ = ratio of specific heats (C_p/C_v) of the gas,
S = internal surface area of the spring,
V = mean volume of the spring,
α = thermal diffusivity of the spring gas $\alpha = D/\rho C_p$,
D = thermal conductivity of the gas,
ρ = mean density of the spring gas,
C_p = specific heat at constant pressure,
ω = circular frequency, $\omega = 2f_n$.

The numerical value of the empirical constant F was determined by Breckenridge *et al* to be 2.6 for the spring used in their rotary stroker cryocooler. This included the enhancement of the heat transfer process due to convective effects as well as the conductive process considered in the basic analysis.

Subsequently, Curwen *et al* (1974), in tests with a gas spring using air and helium, obtained enhancement factors appreciably higher than the 2.6 determined by Breckenridge. Wood (1980b) has reported the experience at Sunpower with many gas springs that a realistic value for the empirical constant is two or three times the Breckenridge value, i.e. F = 6 or 7.

The above equation shows that the losses are proportional to the area to volume ratio (S/V). To minimize the spring loss it is therefore important to minimize the surface area of the spring chamber through the use of spherical shapes or, if cylindrical, to make the length of the cylinder approximately the same as the diameter.

Helium with a high value of the ratio of specific heats γ and thermal diffusivity α results in higher spring losses than gases with low γ and α ratios (Freon 115). However, use of the same fluid for the gas springs and the working space is mandatory in free-piston Stirling engines to eliminate the sealing and charging problems arising from the use of different fluids.

Breckenridge *et al* (1971) briefly discussed several concepts for approaching isothermal operation of gas spring to minimize the hysteresis power losses. So far as is known no investigation of these possibilities was undertaken; readers are urged to re-examine the concepts contained in the Breckenridge proposals.

Two basic types of gas springs are shown in Figure 3.4. The style on the top using the displacer rod as the piston is easy to manufacture and requires less space than the stepped rod design shown on the bottom figure. Wood (1980b) has shown the stepped rod design to be advantageous in reducing gas spring losses due to both hysteresis and leakage.

This loss mechanism is also present in the workspace of the engine as Lee (1981) has noted. Its effect there, however, is greatly enhanced by turbulent mixing which is investigated in detail in Lee and Smith (1980).

Available Programs

The subject of second order analysis is indeed vast, and the discussion above is merely intended to outline the major published approaches and loss mechanisms to consider when choosing a scheme for a particular application. The Martini (1978) and Lee (1981) analysis schemes show promising predictive capability though this is based on a limited number of applications at this time. As more experience is gained with these and other second order analyses, and as more data on actual engine tests and third order predictions become known for comparison, improvements in second order analysis techniques can be expected.

Availability will improve also. In addition to the ones already mentioned, Heames *et al* at Argonne National Laboratory recently developed a computer coded second order analysis package intended for public use; it will be made available through the National Energy Software Center. The Argonne analysis is based on the work of Rios (1969).

Thus it is recommended that one consider the adoption of a ready-made scheme before jumping in to create another as tempting as the latter may be to the analytically inclined. Chances are high at this point that a suitable program already exists which can be used as is or quickly modified to suit the purpose at hand. Of course special needs or novel engine designs may necessitate such extensive modifications to existing programs that starting from scratch would be fully justified.

THIRD ORDER ANALYSIS

As mentioned above, the third level of analysis results when one attempts to directly avoid major idealizations at the basic assumption level. One wishes to more or less thoroughly and interactively model the many simultaneous processes throughout the engine. As idealizations are reduced and fidelity improves, computational complexity rapidly increases. A full capacity digital computer is required to carry out the simulation and processing time is non-negligible even for a single cycle.

So vast is this subject that even less ground can be covered here than was covered for second order analysis. However, the same remarks as made for second order apply here relative to the ability of the seriously interested reader to delve into the literature. Martini (1978), Walker (1980) and Urieli (1979) provide useful surveys of the subject and sound guides to the literature. Furthermore, a recent book by Urieli and Berchowitz (1983) does full justice to the subject. Thus here we shall restrict our discussion to a very brief statement of the fundamentals and a brief indication of how one can best use such analysis programs for the design of free-piston engines in particular.

Ideally, the goal of a third order analysis is to

simultaneously model the energy and fluid flows of the Stirling engine interactively. In a real Stirling engine the working substance does not undergo a single uniform process but rather exhibits a variety of processes from one end of the engine to the other, often with conceptual discontinuities at the natural junctures between sections. Thus most advanced analysis programs partition the engine into a number of regions - variously called cells, control volumes, or nodes - each of which is then treated in a uniform way. Figure 3.6 illustrates a decomposition into 15 cells. This basic starting point was pioneered by Finkelstein and developed in a series of papers to which the (1975) paper is a good entry.

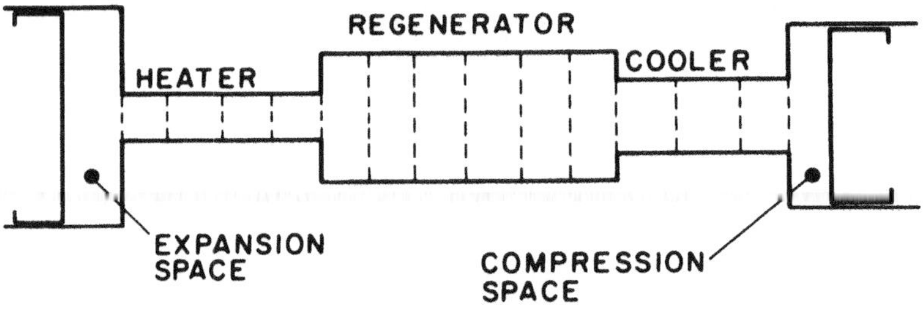

Figure 3.6 Partition of the basic Stirling engine into cells.

Various assumptions are made for the cells regarding fluid behavior and energy exchange so that the fundamental equations (energy, mass, momentum) for the gas can be solved by reasonable numerical methods. For example, the pressure in each cell is assumed constant from one end to the other, and the volume for all cells is constant except for the expansion and compression space volumes. These are usually taken as varying sinusoidally but at this level of analysis can just as easily be made to vary as the machine under consideration dictates; for free-piston machines this means calculating piston and displacer motion from the gas and other forces acting on them. Correlations are chosen to model heat transfer and fluid friction in a way appropriate to the particular control volume location and machine being modelled. Usually the fluid is assumed ideal, but other state equations have been used, for example by Tew *et al* (1978). One chooses at this point whether and how to model other phenomena as suits the purpose. One thus ends up with a system of differential equations indexed by the control volume numbers which can be integrated numerically in time steps to yield the fluid properties as functions of time (and control volume location). From this information overall performance can be easily calculated.

Urieli *et al* (1977) presents a clear and much more detailed description of this type of analysis scheme. Several other control-volume schemes have been developed by making different assumptions relative to the control volumes, the inter-relations among adjacent cells, gas momentum effects, etc. Descriptions of these are to be found in the references cited above. Recently, Organ (1982) and Larson (1982) have suggested using the method of characteristics for third order computer simulations.

Here, even more than with second order analysis programs, the newcomer requiring advanced analysis should seek out a ready-made program. The few that have been already developed and used are in each case the product of an enormous amount of programming and debugging time. In the case of programs that have been used extensively to design real machines (e.g. Sunpower's third order analysis, Gedeon (1978)) the computer code development includes comparing the code to the hardware that it simulates, and making modifications and corrections as necessary. The point is that it takes years of dedicated work to come up with a trustworthy third order computer program. So unless one is willing to devote several years of effort to make one's own from scratch, it is advisable to contract such work to those who have painstakingly developed and nurtured such programs to maturity.

ANALYTIC DESIGN OF FREE-PISTON ENGINES

The complexity of third order analysis schemes makes them ideal for providing very detailed information about what is happening in each major region of the engine. This is invaluable for the designer who wishes to troubleshoot a particular design. For example, it enables one to check, say the heater, for adequate heat transfer or the regenerator for acceptable effectiveness at any desired set of operating conditions (speed, mean pressure, heater temperature, etc.). It provides information (presumably accurate) that is difficult or even impossible to extract from the actual engine itself even if it were available in the lab for testing.

However, these programs are relatively slow running, even with subroutines to accelerate the attainment of steady state operation from start-up. And, therefore, they are costly and somewhat limited for use in the early stages of an engine design project, where a quick user-responsive design guide is needed. Thus it is advisable to have an in-house second order analysis program ever standing by whether or not the higher analysis is done in-house or contracted to a third party. And fortunately, as already remarked, these are becoming more and more available. It is interesting to note that a third order program can often be modified to produce a second order version. For example, one could assume isothermal heater, cooler, and regenerator sections in order to get quick estimates of the pressure wave and flow losses. The advantage of this approach is that one uses the same terminology for all the programs so that information from one can be easily transferred to the other.

Assuming then that one has analytic tools available of an order appropriate for the desired results the basic design problem for Stirling engines is to obtain specifications for an engine which will satisfy power, efficiency, size, fabrication and other requirements in the best way possible. This comes under the general heading of *optimization*. A masterful discussion of this problem was given by Gedeon (1978). The result of an optimization is a set of engine dimensions and operating conditions satisfying the requirements. Of course these specifications may arrive at the designer's desk from less carefully examined sources, raw intuition often contributing much to a new design.

Whatever the source, at some point in the design process the desired strokes and phase angle of the piston and displacer are determined. This is then simply built into the mechanism in the case of a kinematic Stirling engine. But the designer of the free-piston engine is faced with the very difficult task of insuring that the unrestrained engine will in fact operate with the desired frequency, strokes, and phasing under the given conditions. It is fitting to close this chapter with a brief indication of how one can effectively use analysis schemes to solve this fundamental problem for free-piston engines.

Now one could simply specify some of the hardware and then decide the rest by iteration or outright trial and error. For example, if one has only a first order analysis available, then one could estimate or decide on cylinder bore, piston and displacer mass, displacer damping and say piston springing and use the linearized model described above. Displacer springing and piston load damping could be varied until the desired phase and frequency are obtained. But this is rather difficult to carry off from scratch because of the multiple sensitive interactive links between the components. Once one is near the desired point, then perturbation is a very practical approach, but otherwise it is very uncertain. With second or third order analysis programs available, the situation is not much better for this approach because of the longer computation time required; however, more detail is available following a run to guide the new choice of parameters for the next run.

A far better approach is through a *constrained* simulation (Gedeon (1978)). Here the engine is simulated as though the piston and displacer were mechanically constrained to move as desired. One assumes sinusoidal piston and displacer motion with the desired strokes and phasing and analytically simulates the engine running at the frequency and other conditions required. A third order analysis will produce a workspace pressure function at the piston and a pressure drop force function on the displacer. These can be approximated (rather well) by sinusoidal functions (e.g. by expanding into Fourier series and discarding the higher harmonic terms) and thus represented by rotating vectors as described in Chapter 2. It is then a simple matter to find the required piston spring and load damper, and the displacer spring and rod area required. In the case of a second order analysis, the information will be less accurate and much less detailed particularly regarding the displacer damping, but from the estimated pumping power one can easily determine an equivalent damping coefficient and then proceed to the rotating vector diagram.

As a final check, one then can simulate the engine in an unconstrained third order analysis where it is usually found to operate very close to the constrained conditions. From here one would be advised to carry out a collection of unconstrained simulation runs over the range of expected operating conditions to examine the general operational characteristics of the engine, particularly its stability.

NOMENCLATURE

V_e	instantaneous volume of expansion space
V_c	instantaneous volume of compression space
V_D	void volume
p	workspace pressure
p_o	buffer space pressure
p_m	mean workspace pressure
T_E	temperature of expansion space
T_C	temperature of compression space
T_D	temperature of void volume
m	mass of working fluid
R	gas constant
W	cyclic work
P	engine power
f	engine frequency
M, M_p	piston mass
M_D	displacer mass
x_p	piston position from midstroke
x_d	displacer position from midstroke
C_R, C_D, C_P	damping coefficients
K, K_R, K_P, K_D	spring constants
A_P	piston area
A_R	displacer rod area

REFERENCES

Beale, W.T. (1976). Rough Rules of Thumb for Free-Piston Stirling Engine Design. Report, Sunpower, Inc., Athens, Ohio.

Berchowitz, D.M. and Wyatt-Mair, G.F. (1979). Closed Form Solutions for a Coupled Ideal Analysis of Free-Piston Stirling Engines. Research Report No. 79, School of Mechanical Engineering, University of the Witwatersrand, Johannesburg.

Breckenridge, R.W., Heuchling, T.P. and Moore, R.W. (1971). Rotary-Reciprocating Crogenic Refrigeration System Studies. Part I Analysis. Tech. Report AFFDL-TR-71-115, Part I, Wright Patterson Air Force Base, Dayton, Ohio.

Curen, P.W., White, H.V. and Gray, S. (1974). Design and Test of a Helium Gas Spring. Report No. AD-787-142, NTIS, Springfield, Va.

Finkelstein, T. (1960a). Generalized Thermodynamic Analysis of Stirling Engines. S.A.E. Paper No. 118B.

Finkelstein, T. (1960b). Optimization of Phase Angle and Volume Ratios in Stirling Engines. S.A.E. Paper No. 118C.

Finkelstein, T. (1975). Computer Analysis of Stirling Engines. Proc. 10th I.E.C.E.C., Paper No. 759140, pp. 933-941.

Gedeon, D.R. (1978). The Optimization of Stirling Cycle Machines. Proc. 13th I.E.C.E.C., Paper No. 789193, pp. 1784-1790.

Heames, T.J., Daley, J.G., Uherka, D.J. and Zabel, J. (1982). A User Oriented Design System for Stirling Cycle Codes. Proc. 17th I.E.C.E.C., Paper No. 829278, pp. 1681-1687.

Kirkley, D.W. (1962). Determination of the Optimum Configuration for a Stirling Engine. Jour. Mech. Eng. Sci., Vol. 4, No. 3, pp. 203-212.

Larson, V.H. (1982). Computation Techniques and Computer Programs to Analyze Stirling Cycle Engines Using Characteristic Dynamic Energy Equations. Proc. 17th I.E.C.E.C., Paper No. 829283, pp. 1710-1715.

Lee, K. and Smith, J.L. Jr. (1980). Performance Loss Due to Transient Heat Transfer in the Cylinders of Stirling Engines. Proc. 15th I.E.C.E.C.

Lee, K. (1981). Thermodynamic Description of the Adiabatic Second Order Analysis for Stirling Engines. Proc. 16th I.E.C.E.C., Paper No. 819794, pp. 1919-1924.

Martini, W. (1878). Stirling Engine Design Manual. NASA Report No. CR-135382, (NTIS No. N78-23999).

Organ, A.J. (1982). Gas Dynamics of Stirling Cycle Machines. Proc. Stirling Engines Conference, Univ. of Reading, I. Mech. Eng., Paper No. C25/82, pp. 131-140.

Rios, P.A. (1969). An Analytical and Experimental Investigation of the Stirling Cycle. Ph.D. Thesis, MIT.

Rios, P.A. (1971). An Approximate Solution to the Shuttle Heat-Transfer Losses in a Reciprocating Machine. Trans. ASME Journal of Engineering for Power, April.

Schmidt, G. (1871). Theore der Lehmannschen Calorischen Maschine. Z. Verb. dt Ing., Vol. 15, No. 1.

Senft, J.R. (1982a). A Simple Derivation of the Generalized Beale Number. Proc. 17th I.E.C.E.C., Paper No. 829273, pp. 1652-1655.

Senft, J.R. (1982b). Small Stationary Stirling Engine Design. Proc. Stirling Engine Conference, Univ. of Reading, I. Mech. Eng., Paper No. C19/82, pp. 77-84.

Tew, R., Jefferies, K. and Miao, D. (1978). A Stirling Engine Computer Model for Performance Calculations. NASA Report No. TM-78884.

Urieli, I., Rallis, C.J. and Berchowitz, D.M. (1977). Computer Simulation of Stirling Cycle Machines. Proc. 12th I.E.C.E.C., Paper No. 779252, pp. 1512-1521.

Urieli, I. (1979). A Review of Stirling Cycle Machine Analysis. Proc. 14th I.E.C.E.C., Paper No. 799236, pp. 1086-1090.

Urieli, I. and Berchowitz, D.M. (1983). Stirling Cycle Engine Analysis. Adam Hilger Ltd.

Walker, G. (1962). An Optimization of the Principle Design Parameters of Stirling-Cycle Machines. Jour. Mech. Eng. Sci., Vol. 4, No. 3, pp. 226-240.

Walker, G. and Khan, M. (1965). The Theoretical Performance of Stirling-Cycle Machines. S.A.E. Paper No. 949A.

Walker, G. (1979). Elementary Design Guidelines for Stirling Engines. Proc. 14th I.E.C.E.C., Paper No. 799230, pp. 1066-1068.

Walker, G. (1980). Stirling Engines. Oxford University Press, Oxford.

West, C.D. (1980). An Analytical Solution for a Stirling Machine with an Adiabatic Cylinder. Proc. 15th I.E.C.E.C., Paper No. 809453, pp. 2274-2277.

West, C.D. (1981). Theoretical Basis for the Beale Number. Proc. 16th I.E.C.E.C., Paper No. 819787, pp. 1886-1887.

Wood, J.G. (1980a). A Program for Predicting the Dynamics of Free-Piston Stirling Engines. M.Sc. Thesis, Ohio University, Athens, Ohio.

Wood, J.G. (1980b). Lecture Notes for Free-Piston Stirling Engine Workshop. Sunpower Inc., Athens, Ohio.

Wood, J.G. (1982). Private Communication. Sunpower Inc., Athens, Ohio.

CHAPTER 4 - THE SUNPOWER ENGINES

INTRODUCTION

This chapter is basically a catalog with a little history of the Sunpower free-piston Stirling engine. It is the product of a joint effort by Dr. Jim Senft and the founder and president of Sunpower, William Beale. Senft worked at Sunpower for several years before returning to the groves of academe. He is thus uniquely qualified to write with an insider's viewpoint but from a detached position free of the customary commercial pressures.

Sunpower Inc. is a group of enormously talented engineers and scientists that grew, under the leadership of William Beale, out of the Department of Mechanical Engineering, Ohio University, with the objective of developing commercial applications of the free-piston Stirling engines. The group have retained much of the freedom of expression, independent thought, and willingness to share and freely give the benefit of their experience without regard to personal gain, that characterizes the finest in university based research.

The free-piston Stirling engine is a paradox of apparent simplicity that is, in fact, the most difficult of Stirling engines to successfully reduce to practice. The Sunpower group have accumulated more practical experience and have achieved more successes and failures than any group working on Stirling engines.

I recommend that readers interested in free-piston Stirling engine development pay the closest attention to the many lessons that may be learned from the recital of their experience contained therein. I am most grateful to the Sunpower staff, to William Beale and to Dr. Senft for this partial lifting of the veil.

G. Walker

CHRONOLOGY

The free-piston Stirling engine, with its displacer driven by differential areas between the compression and expansion spaces, was conceived by Prof. William T. Beale of Ohio University in March of 1964. The first concept had two pistons and two displacers in parallel cylinders as indicated schematically in Beale's sketch shown in Figure 4.1. A complex and quickly constructed version of this idea was tried in the fall of 1964, but in Beale's own words: "It produced nothing but despair".

Many experiments and trials followed over the next two years, including using an air/water mixture as the working fluid, which gave promising but on the whole erratic performance. Successively simpler versions of the basic idea eventually emerged over this period. Clear

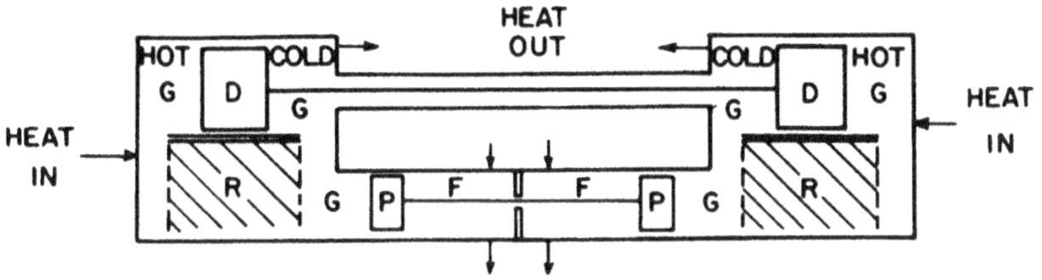

Figure 4.1 The schematic diagram of W. Beale's first free-piston Stirling engine concept.

success came in the summer of 1966; for the first time a free-piston Stirling engine ran with assurance. This engine was extremely simple, used air at atmospheric pressure as the working fluid, had an elementary annular gap regenerator, and relied on close fits for the piston and displacer rod seals. It was a single cylinder engine, positioned vertically to minimize frictional drag, and common rubber bands were used to suspend the piston and displacer against gravity and thereby make self-starting easier.

Inspired by this success the possibilities of the free-piston Stirling were thoroughly explored during the next few years. This was a most creative period for Beale and his students; virtually all of the important applications of free-piston engines that are presently known have roots in this period including the free-piston linear alternator application, free-piston fluid pumps, the duplex Stirling heat pump and the free-cylinder engine.

The free-cylinder engine was the most successful of the early engines. One particularly simple free-cylinder engine operating with helium at a mean pressure of 4 bar was rigged to pump water. This self-starting, electrically heated, engine delivered 70 watts of pumped water with an overall efficiency of 10%. Figure 4.2 is a diagrammatic cross-section of the engine and Table 4.1 contains the principal specifications and performance data. Further details of the major innovations and experiments of this period are to be found in Beale (1969). Work on the free-cylinder water pump concept continued and several models were built and tested; Beale *et al* (1971) describes some of these early results.

Beginning in 1970, some of the simple engines that Beale and his students were building and testing were sold as demonstration units. Known as the "Model 10", about 40 of these units were distributed to private companies, universities and government agencies during the ensuing decade. Figure 4.3 shows an early Model 10 with a linear alternator for the load device, but an interchangeable inertia water pump was also available, With a bore of 32 mm and a stroke of up to 30 mm, these machines operated at a frequency of from 10 to 30 Hz, with Helium as the working fluid at charge pressures up to 5 bar. The engines were available in versions suitable for heating by electricity,

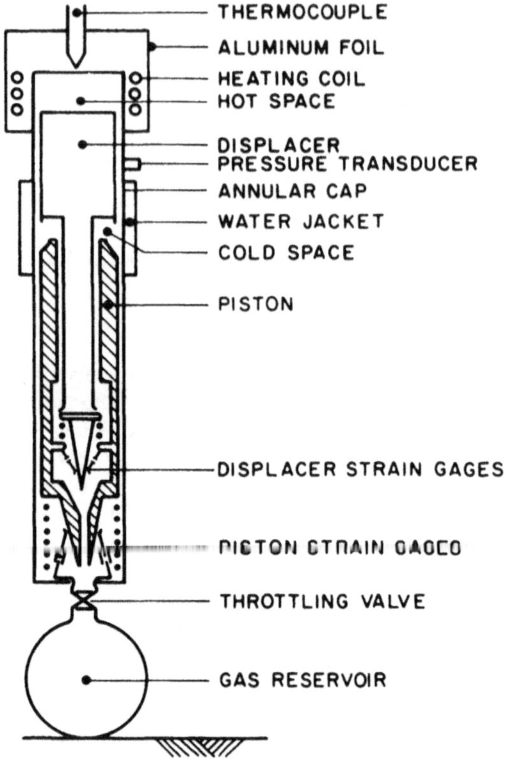

Figure 4.2 A lab engine model with throttling load which was later modified (by closing the end with the throttle valve) to operate successfully as a free-cylinder water pump delivering 70 watts to the water with 5 bar helium charge.

Table 4.1
Principal Characteristics of the Early Beale Free-Cylinder Stirling Engine Water Pump

Displacer Mass	.213 kg
Piston Mass	3.03 kg
Cylinder Bore	51 mm
Displacer Length	150 mm
Piston Length	150 mm
Rod Diameter	19 mm
Displacer/Cylinder Gap	.63 mm
Working Fluid (by volume)	80% He, 20% Air
Total Indicated Work/Cycle	8.8 Joules
Total Rod Work/Cycle	2.0 Joules
Net Indicated Work/Cycle (1/2 Rod Work Assumed Lost)	7.8 Joules
Engine Frequency	12 Hz
Net Indicated Power	94 watts
Power to Heater	940 watts
Indicated Efficiency	10%

Figure 4.3 An early Model 10 demonstrator engine equipped with a linear alternator for electrical output.

propane, or solar energy concentrated by a plastic Fresnel lens. Power output to the load varied between 5 and 20 watts depending upon the heating method, and showed efficiencies of between 5 and 10 percent.

Apart from their value as demonstrators, the Model 10's proved to be of immense value to Sunpower's development program over the years. With several always on hand in the lab, and a generous supply of spare parts lying about, they were often called upon to serve as test beds for new ideas. Being of small and simple design, modifications could usually be carried out quickly and inexpensively. The results were measured on equipment already installed and readily compared with data amassed from the runs of production Model 10's. The time from raw idea to test bed data was often literally a matter of only a day or two. Thus Sunpower had an accurate guide for basic engineering and management decisions.

In 1972, Beale and his students began work on a gas-fired heat pump* development program for the American Gas Association (AGA).

---

*Readers are reminded that in much North American technical literature no clear distinction is made between refrigerating machine or heat pump. Heat pump is the buzz word for the refrigerating unit of an air conditioning system, but gas-fired heat pumps can sometimes be used for either or both heating or cooling applications. G.W.

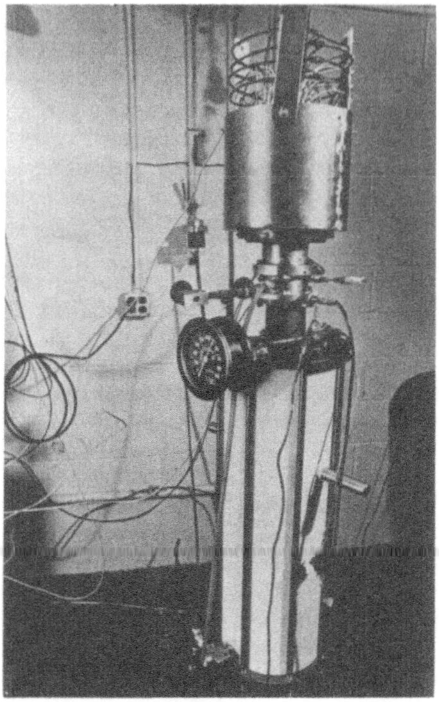

Figure 4.4 The first Stirling/Rankine heat pump in the AGA program was built in 1972. The electrically heated machine ran at a frequency of 20 Hz with an engine power of 1 kW and an efficiency of about 20%.

Figure 4.4 shows the first Stirling/Rankine heat pump built under the program. The basic concept was that of a gas-fired free-piston Stirling engine producing the power to drive an inertia type Freon compressor for a Rankine heat pump system. The inertia pump used on both the AGA machine and the Model 10 water pumps was especially designed to avoid leakage of the pumped fluid into the engine buffer/workspace fluid or vice-versa. In essence, a massive fluid pump piston remained stationary in space while the pump cylinder was reciprocated by the engine piston. The pump piston thus moved relative to the cylinder and so fluid was displaced. No mechanical drive attachment to the pump piston was required, and the whole pump assembly was enclosed in a hermetically sealed pump cylinder with flexible inlet and outlet hoses or tubes. The pump cylinder was attached to the piston of the Beale free-piston Stirling engine and was carried within the bounce/buffer space of the engine. The possibility of fluid contamination was therefore eliminated and furthermore no high friction sliding seals were required. Beale employed helical coils of steel tubing to convey Freon to the pump assembly; one of the coils is clearly visible in Figure 4.5. The buffer space chamber has been removed from the engine assembly of a 'Type II' machine, the second engine/pump design of the

Figure 4.5 A type II AGA engine with the buffer chamber removed showing the inertia freon pump and the helical connecting tube. Engine power was about 2 kW.

AGA program. The inertia pump concept was fully described by Beale *et al* (1973).

Two more machine types were designed and tested under the AGA program, the "type III" and "type IV". The type II and III machines employed multiple-canister style regenerators and coolers whereas the type IV engine design reverted to the simpler annular type of regenerator and cooler used in the original design. The piston rings of the early types were discarded in the type IV engines in favor of hard-surfaced, close-fit, piston and rod seals. All engines in the program employed Helium as the working fluid at a mean pressure of 70 bar. Type IV engines were also sold to Ormat Turbines and Mechanical Technology Incorporated (MTI).

During the initial stages, all of the AGA machines were electrically heated by using the heater tubes as resistance elements. This worked extremely well, allowing the heater temperature to be closely controlled during test runs, and permitted development efforts to be concentrated on the internal engine and pump design. The final engine constructed under AGA sponsorship shown in Figure 6.6 was gas-fired. The Eaton Corp. designed and built the combustion system. Following successful testing of this engine in 1976, it was delivered to General Electric for further development as a pre-production prototype unit under

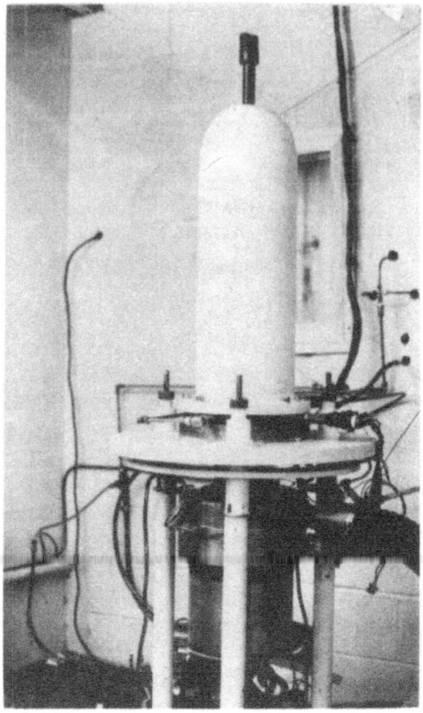

Figure 4.6 A gas-fired type IV machine was built and tested in 1976.

the sponsorship of AGA.

About this time (mid-1970's) Sunpower entered into formal agreement with MTI to work together on a variety of free-piston Stirling engine development projects and studies. The two companies initially became acquainted during the closing stages of the AGA program when MTI was used by Sunpower as a subcontractor for the inertia compressor portion of the heat pump units. Joint projects carried out over the next three years included a design study for a 1 kW space power plant, continued work on gas-fired heat pumps, and solar Stirling alternators.

In this association Sunpower was relied upon for engine analysis. The Sunpower third order computer simulation had been refined over the years by continual development and verification with actual hardware performance in the Sunpower laboratory. With the development of some fast running subsidiary optimization aid programs, by 1978 Sunpower had a highly competent design tool giving good agreement with experiment (Gedeon (1978)).

Much of the hardware design and development work on the joint projects with MTI was done in the Sunpower workshop. A good example of the cooperative activity between Sunpower and MTI is to be found in a 1 kW solar alternator project for the Jet Propulsion

Laboratory. Sunpower assumed the task of designing, building, and testing the engine, while MTI did the alternator. Sunpower started with the type IV AGA engine design, by then highly developed machine, and modified it as necessary to suit the alternator load. The prototype engine was built at Sunpower and fitted with a simple dashpot load for initial testing. A schematic of the machine is shown in Figure 4.7. A significant difference in this engine was that the displacer was

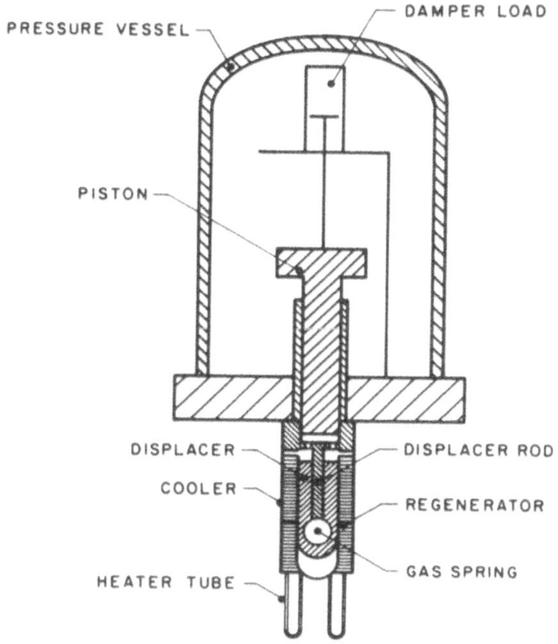

Figure 4.7 A schematic drawing of the JPL engine built at Sunpower in 1978. This engine was based on the AGA heat pump program experience.

sprung to ground rather than to the piston as in the type IV AGA engines. The working fluid was helium at 70 bar pressure. The engine was electrically heated and operated at a nominal 30 Hz producing 1.2 kW at an efficiency of 32%. This prototype engine was run for 50 hours non-stop and delivered to MTI where the alternator was added.

Shortly afterward, a similar engine was built for the NASA Lewis Research Center. A cutaway drawing of this engine, called the RE-1000, is shown in Figure 4.8. The engine delivered 1 kW to the load device at 30 Hz and featured a chrome-oxide surface on the piston and displacer rod running in hardened steel bores.

The small demonstrator engines had by this time evolved into a much improved Model 10-B sold in significant numbers equipped with inertia water pumps, electric generators, inertia compressor freon pumps, and free-cylinder water pumps. Figure 4.9 shows one of these

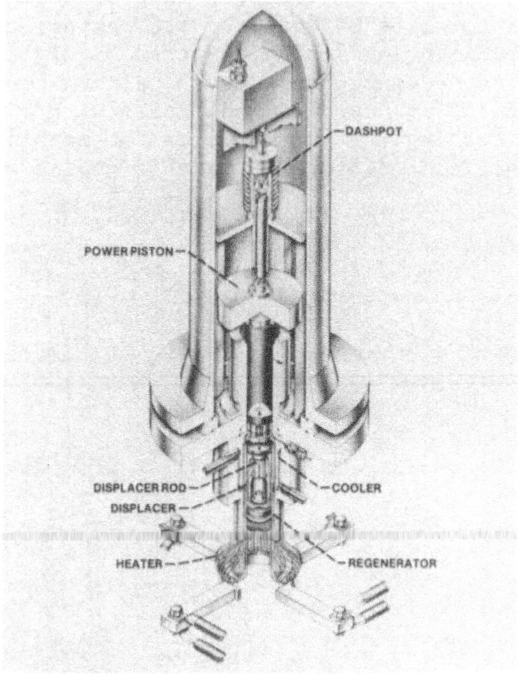

Figure 4.8 A drawing of the Sunpower RE-1000 engine. This engine featured a close fit chrome oxide surfaced piston and displacer rod running in hardened steel bores.

water pumps in operation on the head of a shallow well. The cylinder of the engine was simply attached to the rubber diaphragm of the water pumps, driving it back and forth as the cylinder oscillated. The ring type propane burner moved with the cylinder with a flexible hose connected to the burner to the supply tank. This machine was an excellent self-starter.

In 1978 a development program was initiated within Sunpower to produce a more powerful version of the Model 10. With essentially the same bore and stroke as the Model 10, the new engine was designed for a charge pressure of from 15 to 25 bar, and required extended heat transfer surfaces internally and externally. With these modifications, the engine was capable of driving a linear alternator of 70 to 100 watts output, a significant power increase over the 10 to 20 watts of the Model 10 demonstrators ("demos"). The engine was quickly labeled the "Super Demo", a descriptor retained in suitably reserved style with the company's formal designation for the engine as the SD-100.

A moving coil/permanent magnet field alternator was designed for the power generator version of the new engine, shown in Figure 4.10. A good deal of iron was required for the magnet field, but the outer sleeve served also as the buffer space wall, making a very

Figure 4.9 A free-cylinder version of the Model 10-B driving a diaphragm type water pump.

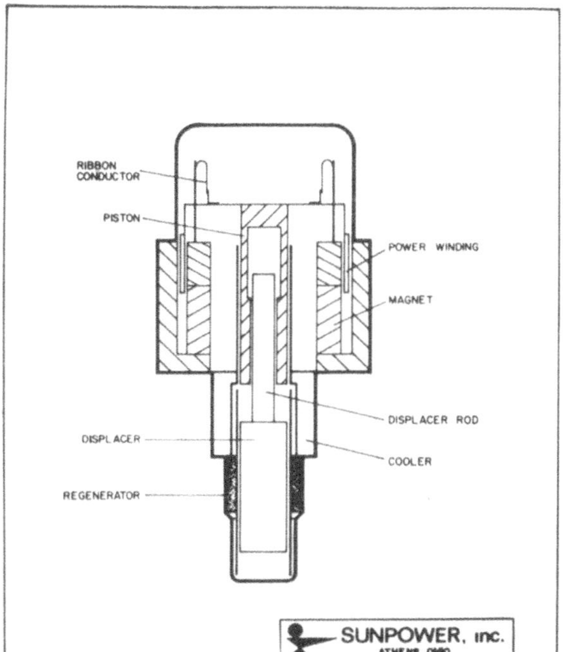

Figure 4.10 A schematic drawing of the first Sunpower "Super Demo" engine which featured the displacer spring in the piston.

compact package. Early models of the engine featured the displacer sprung to the piston, but later versions had a displacer sprung to ground. The 'sprung to piston' arrangement is simpler to construct but with the displacer sprung to ground the displacer is less sensitive to fits, tolerances and wear. The spring to ground is also smaller than the spring to piston, so that the hysteresis loss is reduced. More details of the first SD-100 can be found in Senft (1979).

Figure 4.11 shows a late model SD-100 incorporating several

Figure 4.11 The Model SD-100 free-piston linear alternator unit. This small propane fired power plant delivers up to 100 watts of electricity and is whisper quiet.

features to lower production costs including a helical water cooling tube. Cooling water is circulated by a simple inertia type pump mounted alongside the engine. The engine has a small amplitude of vibration by the isolation spring mount and this is sufficient to operate the water pump.

This Model SD-100 is not yet being mass-produced, although about 15 have been sold (at prototype prices) to parties interested in research and testing for future commercial applications. The 100 watt output of this unit is adequate for various purposes, i.e. operating a color television set and small fan simultaneously yet quiet enough to operate right next to the TV. At sea, the unit can serve as emergency power for navigational and communications equipment, and drive an electric bilge pump as well. Very small vessels can be driven by the SD-100 as illustrated in Figure 4.12. Many other

Figure 4.12 An SD-100 powering a standard trolling motor quietly drives this canoe.

uses for small silent power generators abound and strenuous efforts are being made to mass produce the SD-100.

Beginning in 1978, the Sunpower group have designed, constructed and tested three engines with rotary shaft output. Two were of the full kinematic variety, and therefore fall outside the scope of this work; but further details have been given by Beale (1982) and Wood (1982). The other engine was a hybrid Ringbom-Stirling engine of 1 kW output power, illustrated in Figure 4.13. Air was chosen as the working fluid to avoid seal problems and teflon based materials were used for the non-lubricated piston and rod seals. The piston drive was the common crank/connecting rod type, but a side link (c.f. Senft (1982)) was employed to eliminate piston side loading. Grease-packed sealed rolling element bearings were used in the mechanism for a design life of 2000 hours and the dry crankcase served as the buffer space. The nominal charge pressure of the engine was 7 bar for a 1 kW output at 1000 rpm.

Initially the displacer was equipped with a "tunable" gas spring, but this was later modified to be ineffective except near the extremities of the displacer travel; this concept of "top end tuning" is described in Chapter 5. It is advantageous for obtaining stable operation at low speed while realising efficient operation at higher speeds.

With the introduction of the SD-100, interest was renewed at Sunpower in free-cylinder water pumps. The Model 100 engine was adapted for free-cylinder use and fitted to a variety of diaphragm and induction type pumps. By 1980 several had been built which were capable of pumping useful amounts of water; Figure 4.14 illustrates one of these pumps in action. It is capable of delivering about 30 watts of pumped water ($\sim$ 3 liter meters per second). The engine

Figure 4.13 A 1 kW Ringbom engine built in 1980 at Sunpower.

Figure 4.14 A free-cylinder water pump based on the SD-100 engine reached an output of 40 watts hydraulic power.

operates at a frequency of 25 Hz and a cylinder stroke of from 5 to 20 mm depending upon the head being pumped. Beale (1979) reports further free-cylinder pump concepts.

The SD-100 engine was also adapted for use in the duplex Stirling heat pump shown in Figure 4.15. This simple machine has only three moving parts and is capable of reaching a minimum temperature of 170°K. Three of these units have been built for research groups in

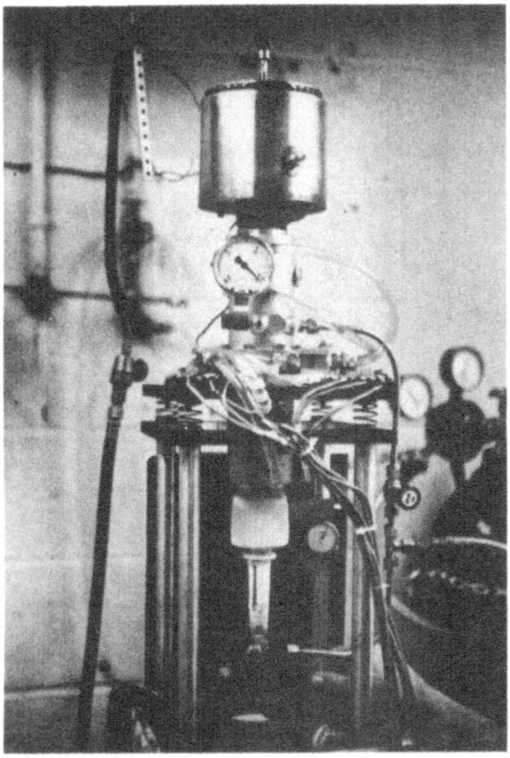

Figure 4.15 A duplex Stirling heat pump based on the SD-100 engine capable of reaching 170 K.

the U.S., Japan, and Europe.

As Sunpower enters the eighties, several new development projects are underway, including larger duplex heating/cooling machines, duplex natural gas liquefiers for use at the well head, a kerosene fired portable power source for remote villages, biomass fired air engines for developing countries, free-cylinder irrigation pumps, and a 10 kW solar electric generator for application in the Near East.

SUNPOWER ADVICE TO THE DESIGNER OF FREE-PISTON STIRLING ENGINES

Years of trial and error, analysis, hopes dashed and hopes sustained have resulted in the following distillation wisdom in the design of free-piston Stirling engines.

1. Adhere closely to Orowan's rule - don't try anything complicated until you have failed at something simple.

2. Start small. Why make a big failure when a small one will teach you as much? But not too small, tiny engines are hard

to make with sufficient precision.

3. To work properly, a free-piston Stirling engine must be really free. Draggy seals and loads that stick can be overcome on crank engines. Free-piston engines should bobble at the slightest touch. Close fits, rather than rings, and cups work best.

4. Seals can be permitted to leak a little as long as they are free to slide - given the choice between a little leak and a little drag, take the leak. In a sealed free-piston engine, a leak is not a loss of working fluid, so much as a loss of power and efficiency.

5. The dynamic and thermodynamic analysis should give the optimum pressure drops across heat exchangers, otherwise either too much power will be lost in pumping, or too much power will be lost in larger temperature drops across heat exchangers. Accurate computer simulation and optimization is essential for good system performance, especially the dynamic analysis in which mass, spring, and damper equations are resolved to estimate the desired relative motions. For this purpose relatively simple thermodynamic relations are adequate to give the approximate pressure-time characteristic.

6. Recognize that free-piston engines are not easy to scale and use the scaling rules (Gedeon (1981)) intelligently.

7. Regard gas springs as a source of high potential losses. They leak and have hysteresis losses. Stiff gas springs can lose lots of power, avoid large Δ V/V.

8. As far as possible make the loading device act more like a damper (velocity dependent force) than a hydraulic pump (fixed force opposing any motion). Easy loads to drive are alternators, heat pumps and inertia coupled fluid pumps.

9. Try to mount the engine so that the axis is vertical to eliminate friction and facilitate starting.

10. Take care to use compatible pairs in sliding fits. Use equal, low thermal efficients of expansion. A hard surface pair, 'hard on hard' is best. Make the fits very close, untapered and round.

11. Use cast heater heads where possible. Brazed tube heater heads give much trouble with fatigue, leaks, cracks and cost.

12. Do not neglect to use the lowly bang-bang (over-driven) displacer. It works when more elegant designs for resonant operation won't.

FREE-PISTON ENGINES AND ROTARY MOTION

There are undoubted advantages to rotating, rather than reciprocating motion - conventional rotary loads may be driven, gearing up or down is facilitated, system cost and weight of rotating machines are often lower than for reciprocating devices, and, not least, there may be more confidence on the part of investors and users in conventional rotating devices rather than unfamiliar reciprocating ones.

For all these reasons there is a temptation to effect a marriage of the hermetically sealed free-piston machine with a rotating load by one means or another. Various forms of union are possible with a greater or lesser degree of practicality. Those considered at Sunpower include:

Hybrid or Ringbom-Stirling Engines - free displacer, crank driven piston - Walker and co-workers at the University of Calgary introduced the idea in 1971. Later in 1980, Senft and Wood at Sunpower designed, built and tested a 1 kW prototype engine.

Hydraulic Output - Rotating Hydraulic Motor - There are many ways to do this, but all are confronted with the seal problem. how to keep the working gas in the engine and the oil out of it. The solution may involve flexing metals, elastomers or sliding seals identical to those used in crank Stirlings, or diaphragms. One configuration tried at Sunpower had a torsion seal comprising a composite structure of elastomer and metal discs that were soft in torsion yet able to sustain large pressure differences. The oscillating shaft drives a gear and sprag clutch combination so that continuous rotary motion is produced. The hermetic seal is retained as well as the lubrication free reciprocating motion of the free-piston engine. The dynamic interaction and variable amplitude of the free-piston engine is also retained. With direct hydraulic pumping action there is the possibility of instability because the engine power output increases with stroke more rapidly than pump power consumption. Distortion of the displacer - piston phase relations, may arise leading again to instability, stalling or displacer/piston collision.

GUIDELINES FOR FREE-PISTON STIRLING ENGINE APPLICATION

1. Do not compete with the internal combustion engine in any of its established areas of dominance. It is too cheap, too effective, too well known, too readily improved to be defeated where diesel fuel is available. The diesel engine is king, don't get in its way.

2. Find new roles not suitable for the internal combustion engine or other established heat engines which the advantage of Stirling engine characteristics, viz.:
   a) heat driven, heat pumps, especially the duplex Stirling-Stirling engine (natural gas liquefier),

b) solar electric generators in bright sun regions,
c) direct biomass fired machines (rice husk fired hot air engine),
d) waste heat utilizers (low temperature engines based on heat pump technology),
e) very simple heat engines less expensive than alternatives (free-cylinder sump pumps).

3. Seek applications that make near-term commercial sense. Avoid exotic applications and materials unless you enjoy R & D for:
   a) its own sake, or
   b) the money the government is willing to squander on it.

REFERENCES

Beale, W.T. (1969). Free-Piston Stirling Engines - Some Model Tests and Simulations. S.A.E. Paper No. 690203, Jan.

Beale, W.T., Rauch, J., Lewis, R. and Mulej, D. (1971). Free-Cylinder Stirling Engines for Solar-Powered Water Pumps. ASME Paper No. 71-WA/Sol-11, Aug.

Beale, W.T., Holmes, W., Lewis, S. and Cheng, E. (1973). Free-Piston Stirling Engines - A Progress Report. S.A.E. Paper No. 730647, June.

Beale, W.T. (1979). A Free-Cylinder Stirling Engine Solar Powered Water Pump. Paper presented at International Solar Energy Society Congress, Atlanta, 1979.

Beale, W.T. (1982). Stirling Engines at Sunpower Incorporated - A Progress Report. Paper C23/82, Institute of Mechanical Engineers Conference on Stirling Engines, University of Reading, March.

Gedeon, D.R. (1978). The Optimization of Stirling Cycle Machines. Paper No. 789193, Proc. 13th I.E.C.E.C., Aug.

Gedeon, D. (1981). Scaling Rules for Stirling Engines. Paper No. 819796, Proc. 16th I.E.C.E.C., Atlanta, Ga.

Senft, J.R. (1979). Advances in Stirling Engine Technology. Proc. 14th I.E.C.E.C., Paper No. 799252, pp. 1175-1179, Aug.

Senft, J.R. (1982). Small Stationary Stirling Engine Design. Paper C19/82, Institute of Mechanical Engineers Conference on Stirling Engines, University of Reading, March.

Wood, J.G., Chagnot, B.J. and Penswick, L.B. (1982). Design of a Low Pressure Air Engine for Third World Use. Proc. 17th I.E.C.E.C., Paper No. 829289, Aug.

CHAPTER 5 - HYBRID OR RINGBOM-STIRLING ENGINES

INTRODUCTION

The Hybrid Concept

A hybrid Stirling engine is a Stirling engine with a free displacer and a crank coupled piston. The term 'hybrid' is a modern one, placing such engines somewhere between the full kinematic and the completely free-piston/free displacer engine. But the concept dates back at least to 1905 when Ossian Ringbom applied for a patent "on a hot air engine in which the movement of the displacing piston is obtained without the connection of rods or cranks or eccentrics or other mechanical parts..."

The Ringbom Engine

The engine of the patent is shown in Figure 5.1. Basically it is a separate cylinder (gamma type) Stirling engine without a mechanical displacer drive. In lieu of the usual links, levers, or cranks, there is only an oversize looking displacer rod with collars.

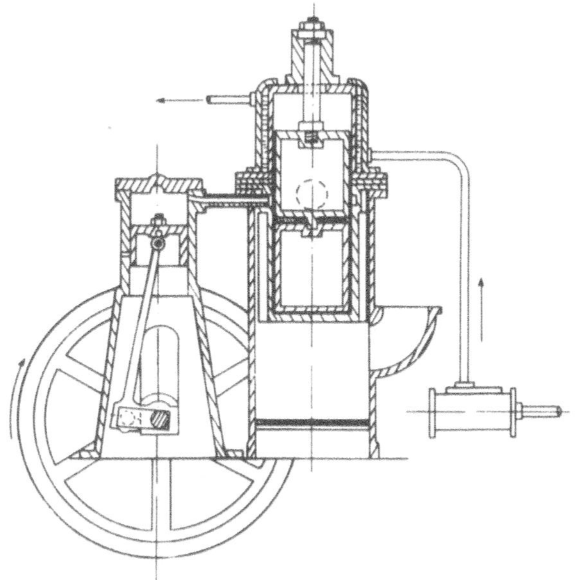

Figure 5.1 Ossian Ringbom's hybrid engine of 1905.

The general operation of the engine will appear clear to anyone familiar with the operation of the completely free Stirling engine. However, as we shall see in this chapter, there are some fundamental differences to be observed to realize optimum operation. A good beginning is an examination of Ringbom's own approach to making a working hybrid.

Starting at the instant depicted in the drawing, flywheel energy will continue to carry the piston upward, compressing the working air within the engine. Near the top of its stroke, the product of the pressure and the rod area generates a force sufficient to lift the displacer to the top of its stroke; assistance is provided from the heating of transferred air during the motion also. To cushion the shock at the ends of the displacer's stroke, Ringbom provided collars at each end of the displacer rod and closely mating sockets at each end of the rod guide. As the collar enters the socket a pneumatic cushion or dashpot is formed to decelerate the displacer.

With the displacer in its full upward position, pressure in the machine is virtually at its peak as the piston begins its downward expansion stroke. The pressure increase due to a higher mean temperature because the displacer is in the uppermost position will maintain the displacer there past the crank angle where the displacer first began to lift. This is an important point for it provides a phase shift additional to that arising from displacer inertia. Still more phase shift results from the socket and collar, but Ringbom is unclear here whether this is due to the larger effective diameter of the rod or to a simple dashpot effect; of course it all depends upon the degree of fit between the collar and the socket. The sum total is a phase shift sufficient to allow much of the piston expansion stroke with the displacer in the uppermost position.

Somewhere before piston bottom dead center, the displacer falls back to its original position, again cushioned by a collar and socket. The patent description attributes gravity as the principal force returning the displacer. If one assumes there was no leakage of air from the workspace during the cycle, then gravity is the only force available because the small port which Ringbom located to open at piston bottom dead center would adjust the minimum cycle pressure to atmospheric. With leakage during the earlier portion of the cycle there would be some pressure assist for the return, but we can safely assume that all attempts were made to minimize leakage.

This then is the Stirling Ringbom hybrid cycle. Reliance upon gravity for displacer return imposes a serious speed limitation on the engine. Far greater acceleration forces are available from gas pressure differences acting on the displacer rod area, or from external springs. Nevertheless, the engine described by Ringbom in the patent probably compared well with the other Stirling engines of the period in performance, and certainly had the advantage of lower manufacturing cost.

OVERDRIVEN MODE OPERATION

Comparison with Free-Piston Engines

It is helpful to begin our discussion of the modern Ringbom-Stirling engine by considering the exact relationship between it and the free-piston Stirling engine. The step from a free-piston Stirling engine to the Ringbom hybrid is not a conceptually difficult one to make; the chronology of these two inventions, however, was the reverse and the inventive steps were much more difficult.

Visualize first a free-piston Stirling operating under steady state conditions, with a carefully tuned spring-mass-damper displacer executing its motion ahead of the piston. The piston, connected to the load device (e.g. an alternator armature or a hydraulic pump), moves in near harmonic motion at constant frequency and amplitude. The idea of replacing the linear load device by a rotary load coupled to the piston by a crank/connecting rod giving the same piston stroke is a relatively easy step. Then with the load adjusted to be equivalent to the one it replaced, the engine will run exactly as before - and the Ringbom principle is reborn.

Today the Ringbom hybrid is closely identified with free-piston Stirling engines and the tendency exists to design and analyze it as one would free-piston Stirlings. However, a little more thought will show there are significant differences. The free-piston machine is essentially a constant frequency machine. It is desirable and customary, therefore, to design the displacer drive system (rod area and external spring) to operate at resonance. As load changes in a free-piston Stirling engine, the piston amplitude changes but the frequency remains more or less constant.

With a crank on the piston, the piston moves with a constant amplitude but a variable frequency in response to changes in loads. Hence with a frequency-tuned displacer, its amplitude and phasing can change drastically as piston frequency changes. In fact, it has been observed that many attempted hybrids will only run over a very narrow speed/load range; small changes stop or even reverse the engine.

Basic Approaches to Design of Hybrids

In light of these major differences one must adopt an alternate strategy to the design of hybrid Ringbom-Stirling engines. A possible approach involves the use of a spring of variable stiffness on the displacer so that the resonant frequency of the displacer can be matched to the shaft speed. A gas spring would be most convenient here for the spring stiffness can be varied simply by changing its mean volume.

The system could probably be made self-governing. Automatic controls would adjust the displacer spring in response to changes in engine speed, load, charge pressure, heater and cooler temperatures, etc.

However, the control system could become quite complicated and thereby negate the principal attraction of the Ringbom arrangement, the simplicity of the displacer drive. Fortunately, an alternative approach is possible that retains the simplicity of Ringbom's engine yet permits stable multispeed operation to high speeds. It involves first eliminating the piston port so that the displacer is driven by gas pressure in *both* directions (gravity can be ignored in fact for high speed engines). Second it requires proportioning the engine components (displacer mass, rod area, piston swept volume, etc.) so that at all desired running speeds the displacer is positively *driven against stops* by gas pressure forces; this is called overdriven mode operation.

The Overdriven Mode Operation Concept

The important principle of this mode of operation is that the displacer is *overdriven* against its stops - overdriven in the sense that it is made to contact each stop *before* the gas pressure difference across the displacer rod reverses to accelerate the displacer in the opposite direction. The motion of the displacer is not merely truncated by physical travel limits but is truncated while experiencing positive acceleration with respect to its direction of motion.

Under these conditions, displacer motion will appear as in Figure 5.2. Prior to point A the displacer is at rest against its

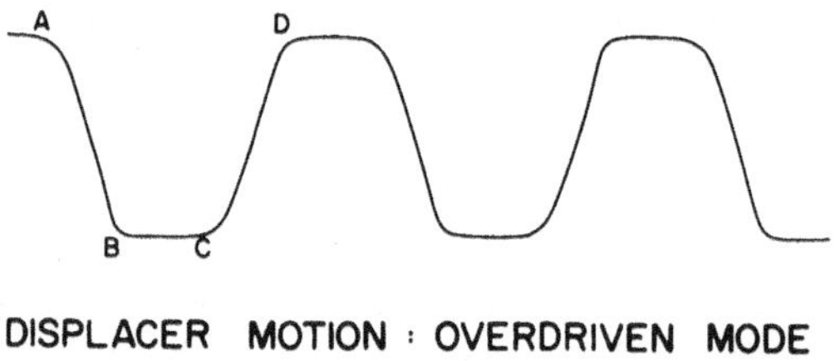

Figure 5.2 Overdriven mode displacer motion.

stop in the uppermost position. The changing gas pressure in the working space begins to move the displacer down until at B the displacer contacts the opposite stop and rapidly halts in the lowermost position. Note the small radius of curvature at B. Dissipative cushions at the travel extremes are assumed to eliminate bounce. The displacer is held motionless against the lower stop by gas pressure forces until point C is reached when the return to the uppermost position begins. The return motion is similar since we have symmetrically varying gas pressure with the piston port omitted.

Although minor bouncing would not seriously affect operations we shall for simplicity assume there is no bounce. Likewise we assume that displacer motion ceases immediately upon contact with its stops. The idealized motion then appears as in Figure 5.3 with

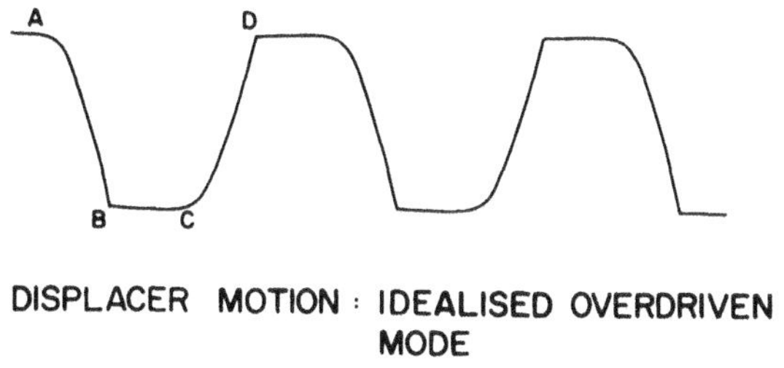

Figure 5.3 Idealized overdriven displacer motion.

sharp corners at points B, D, etc. instead of the rounded curves of the previous figure.

This is the displacer motion which characterizes the *over-driven mode* operation of the hybrid Ringbom-Stirling engine. It will be demonstrated below, that for the entire speed range of overdriven mode operation, the engine will run stably in a steady state with a constant piston/displacer phase angle and produce positive work available at the crankshaft.

Stable Operation

With the usual assumptions of a long connecting rod and a large flywheel we may consider the piston motion of a hybrid to be purely sinusoidal. The piston and displacer motion in the overdriven mode would then appear as in Figure 5.4. As usual the vertical distance between the two curves represents the compression space volume and the distance from the upper horizontal broken line to the displacer curve represents the expansion space volume. We assume steady state operation so that both curves are periodic of the same period.

At $t = 0$, the displacer is at the hot end and the piston is at midstroke. We assume that the pressure in the workspace equals the charge pressure when both piston and displacer are at midstroke. Thus at $t = 0$, the pressure difference across the displacer rod holds the displacer at the hot end.

As the piston moves inward, a point t_0 is reached where the pressure difference across the displacer rod is zero, the point of instability of the displacer. As the piston continues ascending the working space pressure increases and the displacer begins to move down toward the cold end; its acceleration at any time is a function of piston position and its own position. The displacer reaches the limit of its travel at a time t_1, and thence remains at the cold end until, at t_2, it begins the return ascent, exactly half a period from t_0. Although far from ideal, the phasing results in positive work output.

The cycle changes as piston frequency increases. The point t_0, where displacer motion begins, moves toward the left and the

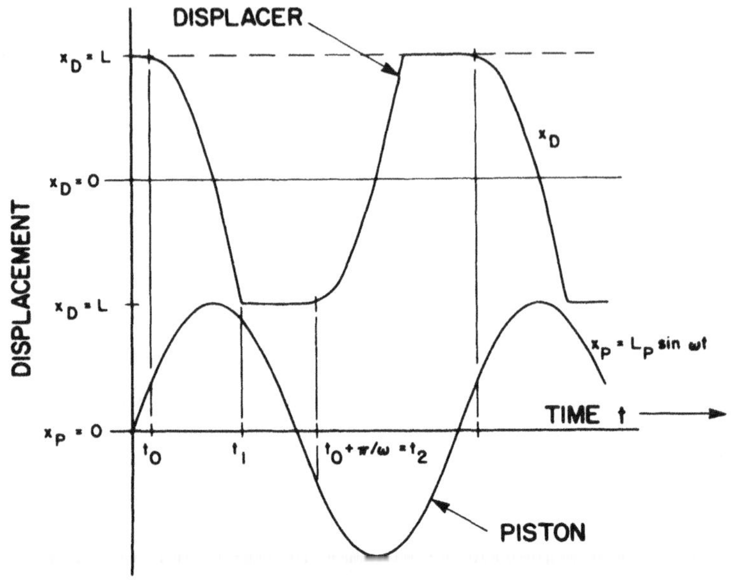

Figure 5.4 Piston and displacer motions in an overdriven hybrid.

transit time, $t_1 - t_0$, decreases as frequency increases also. A phase shift to the left might therefore be expected despite the decrease in the piston period. At the lower frequencies, this may occur but at higher speeds there is a definite shift to the right. For clearly there is a *positive* lower bound for the transit time, namely the transit time one would observe if the piston were moved completely in before the displacer is released, and held in while the displacer completes its motion.

At the higher frequencies, the machine will be operating some-somewhat near this limiting transit time. Let us suppose for the moment that Figure 5.4 represents a high engine speed so that the transit time there is very near minimal. Large increases in frequency then will result in progressive phase shifts toward the right as Figure 5.5(a) and 5.5(b) indicate. This occurs because the transit time does not decrease significantly whereas piston period does. The cycle improves. This continues until, as in Figure 5.5(c) overdriven mode operation is about to cease. Further increases in speed will usually result in constantly changing phasing of the piston and displacer and erratic running; in this way the machine is prevented from over-speeding. This is an extremely valuable characteristic of Ringbom-Stirling engines for it indicates the engines are physically incapable of runaway over-speeding.

FIRST ORDER ANALYSIS

A simple analytic model for Ringbom type engines is presented

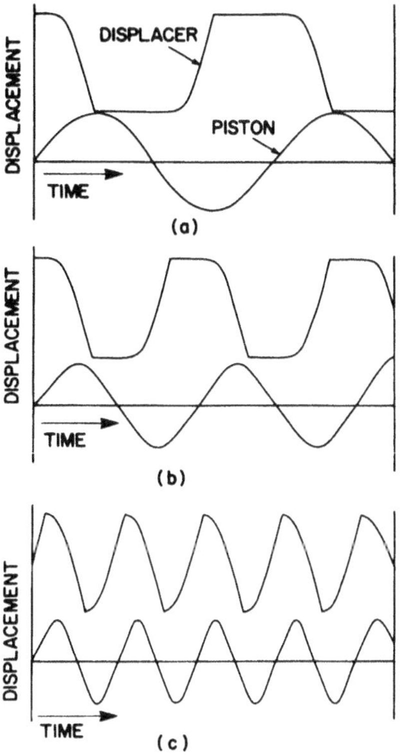

Figure 5.5 Effect of engine speed on displacer motion.

here as described in Senft (1981a). This initial model and its consequences have since been extended in various directions which are indicated at appropriate places below; full references are given for the reader interested in more detail. The principal result (Theorem 5.11) defines a necessary and sufficient condition for stable (overdriven mode) running of these engines. This condition promises to be a useful tool for the preliminary design and for evaluation of free-displacer/crank piston engines. It allows the designer to rapidly choose or modify the principal engine parameters (e.g. rod area, displacer mass, piston stroke, etc.) to obtain the speed range and phasing desired. One can quickly rough-out a design before refining it with more elaborate computer simulations incorporating heat transfer and fluid flow modelling.

Isothermal Model for Hybrid Ringbom-Stirling Engines

Assume the engine configuration shown in Figure 5.6. Arrows in the figure indicate the positive directions taken for the displacer and piston position variables x_D and x_P, respectively. For simplicity we assume isothermal expansion and compression and a uniform pressure

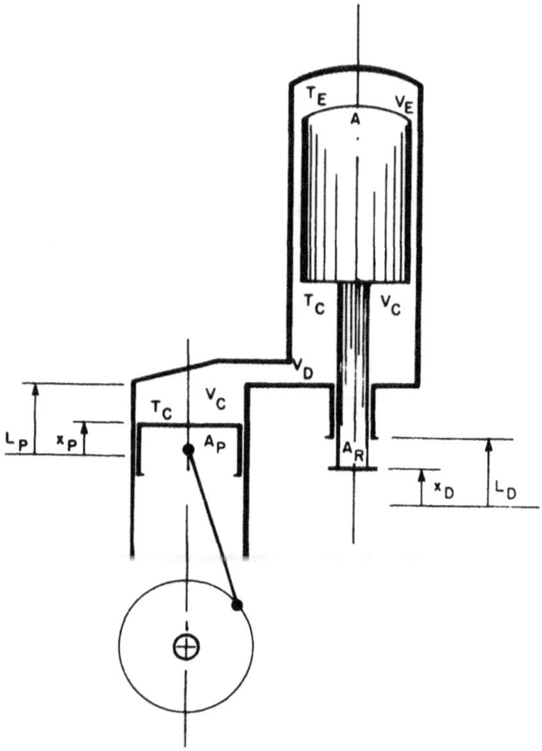

Figure 5.6 Schematic representation of the hybrid Stirling engine.

throughout the work space including a dead space of volume V_D and temperature T_D.

The instantaneous expansion and compression space volumes are given by:

$$V_e = A(L-x_D)$$

and

$$V_c = A(L+x_D) + A_P(L_P-x_P)$$

The volume occupied by the displacer rod is ignored because it is usually insignificant; this is especially true for high pressure engines. Where the rod area A_R is large, its effect can be included with modification to the given equations (Senft (1981b), (1982b), (1983)).

Applying the ideal gas law, $pV = MRT$, to the expansion, compression, and dead spaces yields

$$p = \frac{MRT_C}{A(1-\tau)x_D - A_P x_P + AL(1+\tau+\lambda\kappa+\sigma)} = p(x_D,x_P)$$

The pressure difference across the displacer rod is:

$$\Delta p = p - p_m = p(x_D, x_P) - p(0,0) = \Delta p(x_D, x_P)$$

In order to make the analysis manageable, we replace the above expression for Δp by a linear approximation at (0,0), specifically the differential of p at (0,0) and so find:

$$\Delta p(x_D, x_P) = C_o(\tau - 1)x_D + C_o \kappa \, x_P \qquad (5.1)$$

where

$$C_o = \frac{MRT_C}{AL^2(1+\tau+\lambda\kappa+\sigma)^2}$$

For simplicity the viscous forces acting on the displacer are ignored and we further neglect the forces arising from external springs or dampers acting on the displacer. Therefore, the only force acting on the displacer is:

$$A_R \, \Delta p \, (x_D, x_P)$$

Assuming that the piston motion is purely sinusoidal:

$$x_P = L_p \sin \omega t$$

we obtain the differential equation for displacer motion:

$$\ddot{x}_D - Kx_D = K \left(\frac{\kappa}{\tau - 1}\right) x_P \qquad (5.2)$$

where

$$K = \frac{A_R}{M_D} C_o \; (1 - \tau)$$

The general solution of Equation (5.2) is:

$$x_D = a \exp(\sqrt{K}t) + b \exp(-\sqrt{K}t) + \frac{\lambda\kappa}{(1-\tau)} \frac{KL}{(K+\omega^2)} \sin \omega t \qquad (5.3)$$

where a and b are constants to be determined by initial conditions.

Equation (5.2) applies (and hence (5.3) describes the engine operation) only when the displacer is acted upon solely by forces due to a gas pressure difference Δp across the displacer rod. It does not apply when the displacer is held against a stop, for the stop supplies a force not included in the analysis above. Thus Equation (5.3) is meaningful only when the displacer is in motion.

With the displacer initially at $x_D = L$, Equation (5.2) begins to apply at the point t_0 where $\Delta p = 0$, as shown on Figure 5.4. We find:

$$t_0 = \frac{1}{\omega} \sin^{-1} \left(\frac{1-\tau}{\kappa\lambda}\right) \tag{5.4}$$

Here we see the inequality:

$$\frac{1-\tau}{\kappa\lambda} < 1 \tag{5.5}$$

must be satisfied to initiate displacer motion. From symmetry of the Δp function, we see the identical situation will prevail at $t = t_0 + \pi/\omega$, with the signs reversed, provided the displacer reaches $x_D = -L$ *before* this time. Hence the overdriven mode of operation will occur only if:

from $x_D = L$ and $\dot{x}_D = 0$ at $t = t_0$ the displacer reaches

$$x_D = -L \text{ at a time } t_1 \leq t_0 + \pi/\omega \tag{5.6}$$

The initial conditions $x_D = L$ and $\dot{x}_D = 0$ at $t = t_0$ allow the determination of the constants in solution (5.3). Evaluating a and b, and then introducing the hyperbolic functions we arrive at the desired particular solution of (5.2):

$$x_D = \frac{L\omega^2}{K+\omega^2} \left[\cosh (\sqrt{K}\,(t-t_0)) - \gamma\chi \sinh (\sqrt{K}\,(t-t_0)) + \beta\gamma^2 \sin \omega t\right] \tag{5.7}$$

where

$$\gamma = \sqrt{K}/\omega\,, \qquad \beta = \frac{\lambda\kappa}{1-\tau}\,, \qquad \text{and} \qquad \chi = \sqrt{\beta 2-1}$$

A Criterion for Overdriven Mode Operation

It is virtually impossible to solve (5.7) analytically for t in terms of x_D but it can be done numerically to determine t_1 for use in criterion (5.6). As mentioned in the introduction, a more convenient test for overdriven mode operation, namely Theorem (5.11) below, may be deduced from the following three observations:

i) If the engine operates in the overdriven mode, then $\gamma\chi > 1$ (5.8)

ii) If $\gamma\chi > 1$, overdriven operation occurs if and only if: $\gamma\chi \sinh (\pi\gamma) - \cosh (\pi\gamma) \geq 1$ (5.9)

iii) If $\gamma\chi \sinh (\pi\gamma) - \cosh (\pi\gamma) \geq 1$, then $\gamma\chi > 1$ (5.10)

The deduction of these statements is given in Senft (1981c) and (1982b).

Overdriven Mode Theorem

The above facts imply the following basic theorem for overdriven mode operation of hybrid engines:

> A free-displacer/crank-piston hybrid Ringbom-Stirling engine operates in the overdriven mode if, and only if:

$$\sinh(\pi\gamma) - \cosh(\pi\gamma) \geq 1 \tag{5.11}$$

Alternate forms of the theorem are given in Senft (1983).

Application of the Theorem

The criterion of Theorem (5.11) is particularly convenient for determining the stable running speed range for a proposed hybrid engine design. In the first place, only two dimensionless variables χ and γ are involved. It is easy to solve inequality (5.11) for χ to obtain the equation:

$$\chi = \frac{1 + \cosh(\pi\gamma)}{\gamma \sinh(\pi\gamma)} \tag{5.12}$$

as a description of the boundary of the overdriven mode operation region of the γ-χ plane; Figure 5.7 illustrates this region. In the second place, only γ contains the shaft angular velocity variable ω, and includes it in convenient fashion as a divisor.

By way of example, we can apply (5.11) to the miniature Ringbom-Stirling engine described in Senft (1982a) and shown in Figure 5.8. Table 5.1 gives the engine specifications (temperatures T_E and T_C are estimated, and T_D is taken as the arithmetic average of T_E and T_C), and lists the calculated parameters (SI units are used for dimensional variables).

Solving Equation (5.12) numerically for the value of γ corresponding to the above value of $\chi = .92$, we find $\gamma = 1.15$. Hence by Theorem 5.11, the engine will operate in the overdriven mode if and only if:

$$1.15 \leq \gamma = \frac{\sqrt{K}}{\omega}$$

that is, if and only if:

$$\omega \leq \frac{\sqrt{K}}{1.15} = 77 \text{ rad/sec} \approx 740 \text{ rpm}$$

a deduction which agrees well with actual observation of this miniature

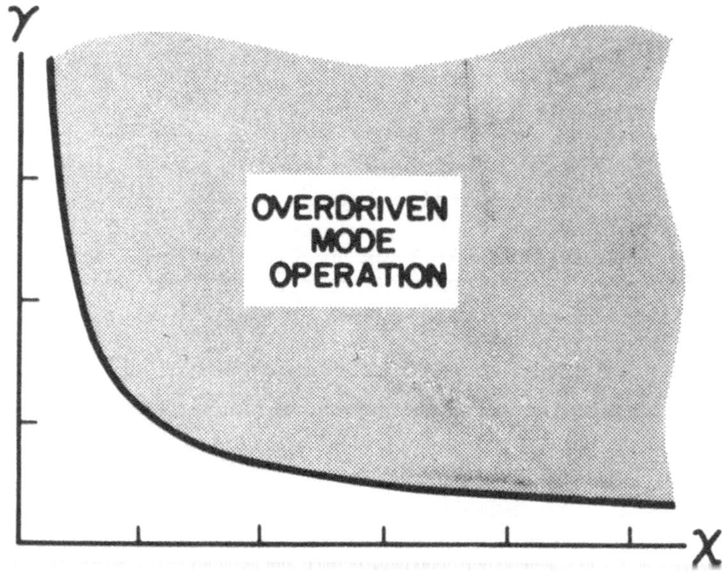

Figure 5.7 The overdriven mode region of the γ-χ plane.

engine. Similar use could be made of (5.11) and (5.12) to determine A_R or M_D from the other parameters and a selected value for $\omega = \omega_{max}$. More details on like uses of the theorem can be found in the references.

SOME PRACTICAL CONSIDERATIONS

Displacer Drive Loss

The overdriven mode concept provides a convenient conceptual model for stable running Ringbom-Stirling engines. As with any such model, one must exercise care when attempting to reduce it to practice.

The major penalty for the simplicity of the Ringbom-Stirling is the energy loss due to displacer motion truncation in the overdriven mode. In the conventional kinematic engine the energy involved in displacer deceleration is transferred to the flywheel, with some loss to friction and is recycled for displacer acceleration, again with some frictional loss in the linkage bearings. In the free-piston Stirling engine having a resonant tuned displacer, the displacer spring plays the same role as the flywheel, alternately accepting and releasing displacer energy with only minor hysteresis losses.

The Ringbom-Stirling engine running in the ideal overdriven mode experiences the loss of all the displacer kinetic energy at the end of each stroke. Of course, one could devise various mechanisms to recover and redirect some of the energy to the flywheel, but at the cost

Figure 5.8 A miniature hybrid engine described in Senft (1982a).

Table 5.1
Specifications of the Ringbom-Stirling Engine of Figure 5.8

Principal Dimensions (see Fig. 5.6)	Calculated Parameter
$A = 1.98\ cm^2$	$\tau = .53$
$A_P = 1.27\ cm^2$	$\tau' = .70$
$A_R = 0.24\ cm^2$	$\sigma = .72$
$L = 0.79\ cm$	$\lambda = 1.00$
$L_P = 0.79\ cm$	$\kappa = .64$
$V_D = 1.6\ cm^3$	$\beta = 1.36$
$M_D = 6.3\ gm$	$\chi = .92$
$P_C = 1\ atm$	$C_o = 4.42\ E6$
$T_C = 400°K$	$K = 7.91\ E3$
$T_E = 750°K$	
$T_D = 575°K$	

of complicating this simplest of rotary shaft output Stirling engines. Fortunately there is a simpler alternative requiring no major physical or constructional complications to avoid substantial displacer energy loss and at the same time allow operation beyond the overdriven range.

This is called top end tuning and is discussed below.

Top End Tuning

An important extension of the overdriven concept called *top end tuning* is realised by employing stiff but elastic cushions instead of purely dissipative stops. The stiffness must be carefully chosen to return energy to the displacer at a rate matching engine speed just beyond the overdriven mode range. Below this speed some bounce will occur but will be minimal and not greatly affect stable overdriven running. As the engine approaches the *design* speed, displacer kinetic energy loss will significantly drop, engine power sharply rise, and smooth running continue above the overdriven speed limit. With top end tuning one can have a stable running engine from low speeds up to the efficient design speed. At this level of design a sophisticated third order simulation programme is virtually essential.

Pneumatic or gas springs provide the ideal cushion for top end tuning since 'contact' losses are small and long life can be realised. A configuration much like Ringbom's collar and socket idea, or its equivalent variant shown in Figure 2.63 could be used if a sufficiently close fit could be achieved. One needs a collar/socket clearance small enough so that the device acts more like a spring than a damper, but as clearance decreases, the 'entry' problem increases. Bevelled leading edges only ensure eventual entry, but do not prevent heavy impacts and surface damage from 'off-center' approaches'. The entry problem can be avoided with a continuous ported cylinder as shown in Figure 5.9. A 'floating' spring piston would be advisable to minimize high tolerance requirements.

General Design Guidelines

Even without top end tuning, losses need not be prohibitively high. Designs for air engines of 1.8 kW indicated power have been prepared with a displacer kinetic energy loss less than 10 percent of the indicated power. This is reckoned to be an acceptable price for the low manufacturing cost that a hybrid makes possible. Moreover, the loss may be substantially decreased by application of simple design guidelines readily deduced from the preceeding sections as follows:

Design for low engine speeds

Choose higher charge pressure or a larger swept volume rather than higher speed to achieve the desired power levels.

Minimize displacer mass

This is important in all engine types but is crucial for Ringbom-Stirling engines. Vent displacer interior to mean pressure to ease strength requirements.

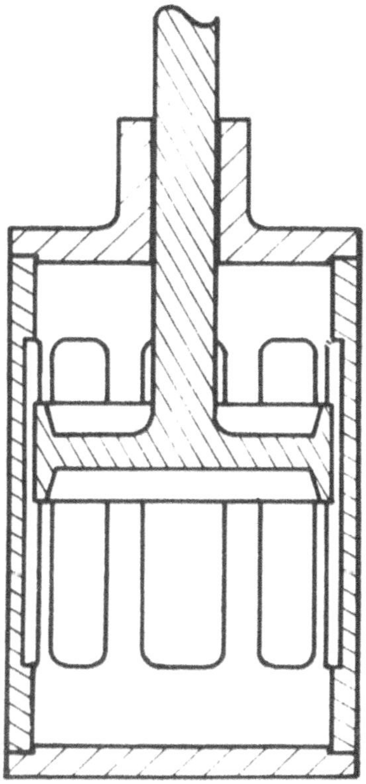

Figure 5.9 Gas spring cushion for top end tuning hybrids.

Adopt a short displacer stroke

Increase the displacer cylinder bore rather than displacer stroke to achieve the desired displacer swept volume.

All of the above tend to decrease displacer kinetic energy loss. They are not completely independent and discretion must be exercised in their application.

PRACTICAL RINGBOM-STIRLING ENGINES

Nothing is known of Ringbom's engines outside of the description contained in the patent (Ringbom (1907)). No subsequent developments were reported in the literature and so far as is known the principle was lost until rediscovered in 1970 by Walker and his colleagues at the University of Calgary, William Beale, Jack Roberts and John Kentfield (Walker (1973)).

They conceived a tandem arrangement, shown in Figure 2.20 with the displacer and piston operating in a single cylinder. The concept was attractive as the means to convert existing internal combustion engines to operate as Stirling cycle machines and hence reduce the development cost of large engines.

A prototype engine was made, as shown in Figure 2.21, based on the conversion of a single cylinder Honda industrial engine. The performance of this unit was disappointing. It was barely self-sustaining and operated at a very low speed. It was designed largely on an intuitive basis and no significant development effort was invested to improve the performance.

About a decade later, the small model Ringbom shown in Figure 5.8 was built to illustrate that stable running high-speed hybrid engines were possible through the concept of overdriven mode operation (Senft (1982a)). As it turned out, this tiny 2 cc machine served to inspire a good deal of full-size Ringbom activity. The first result was the engine by Senft and Wood at Sunpower shown in Figure 4.13. This moderately pressurized air engine had a gas spring top end tuner designed with the aid of the Sunpower third order analysis code. It successfully produced its design output of 1 kW at about 1000 rpm in laboratory tests. These developments in turn rearoused the interest of Walker in the conversion of large locomotive diesel engines to coal-burning engines (Walker (1982)). The possibilities inherent in this work led to a substantial research development contract between the Transportation Development Center, Government of Canada and the University of Calgary, Alberta, Canada, for the provision of a coal-fired Ringbom-Stirling engine based on a diesel conversion. The engine selected for conversion was a single-cylinder Lister diesel engine producing 5.9 kW (8 horsepower) at 14 Hz (850 revolutions per minute) with a bore and stroke of 11.4 cm (4 1/2 ins.) and 13.3 cm (5 1/4 ins.) respectively. A cross-section of the engine and a photograph of the test bed installation with propane gas heating is shown in Figures 5.10 and 5.11, respectively. Air is the working fluid of the engine and water is the lubricant. Water is less effective than oil as a lubricant but is preferred for Stirling air engines to eliminate the risk of explosion with oil leaking into the working space and combining with air in the hot parts of the engine. A further advantage is that water does not accumulate and decompose (carbonize) to block the regenerator as oil does. It accumulates in the cold spaces (the compression space) of the engine. Advantage may be taken of this to convert the unlubricated dry rubbing pneumatic seals into lubricated hydraulic seals to solve the difficult sealing problems of all Stirling engines including the Ringbom-Stirling variety.

Subsequent studies have resulted in new concepts for large coal-fired Stirling engines for marine propulsion, stationary power, locomotives and heavy off-highway vehicles for mining, forestry, construction and agriculture.

One concept for a marine engine is shown in Figure 5.12. This has a large double-acting piston with the crank mechanism contained within the piston and operating as a Scotch yoke. The two end spaces

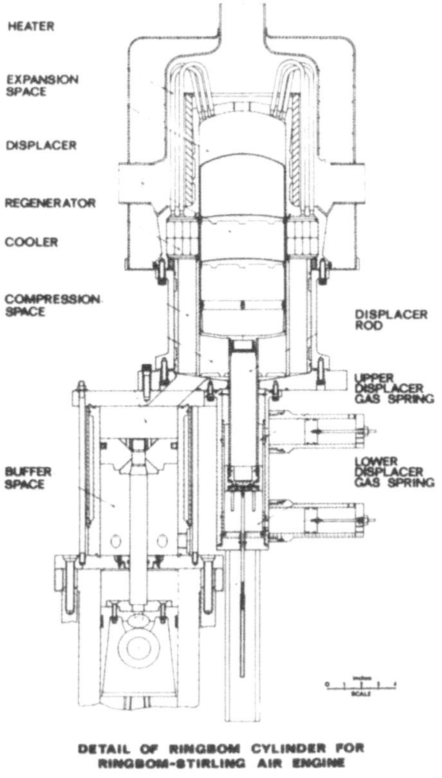

Figure 5.10 Ringbom-Stirling engine based on the conversion of a Lister diesel engine.

of the cylinder containing the piston are coupled to twin displacers. The two displacers are independently driven by an electric motor after the manner devised by Martini. This facilitates starting and reversing of the engine. The displacer drive system also incorporates a Ringbom type displacer drive so that at operation near the design point the power input to the independent displacer drive motor would be reduced to a largely monitoring and regulatory role. Walker *et al* (1983) applied this to locomotive and stationary power applications.

Another recent Ringbom development is the construction and successful operation of an overdriven single-cylinder Ringbom at the University of Wisconsin in River Falls. The engine was designed and built to demonstrate that overdriven mode operation is possible for the single-cylinder arrangement. This of course means that stable relatively high speed operation can be realised for this type of machine.

Figure 5.13 shows the ideal piston and displacer motion of the overdriven single-cylinder Ringbom. It differs from the separate cylinder case of Figure 5.4 in that the piston and displacer travel

Figure 5.11 Ringbom-Stirling engine based on the conversion of a Lister diesel engine on test.

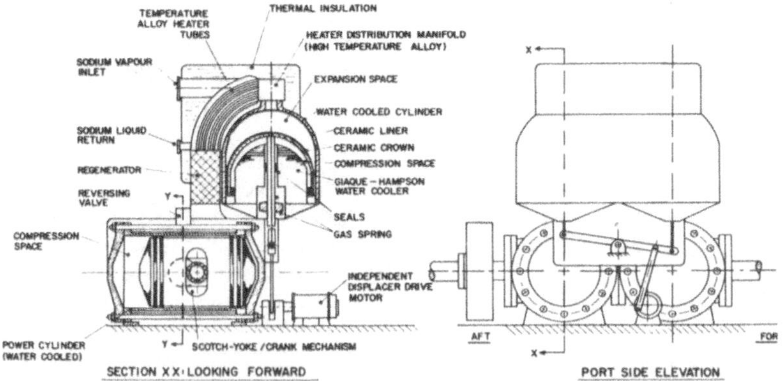

Figure 5.12 Concept for a Martini-Ringbom Stirling engine for marine propulsion or stationary power.

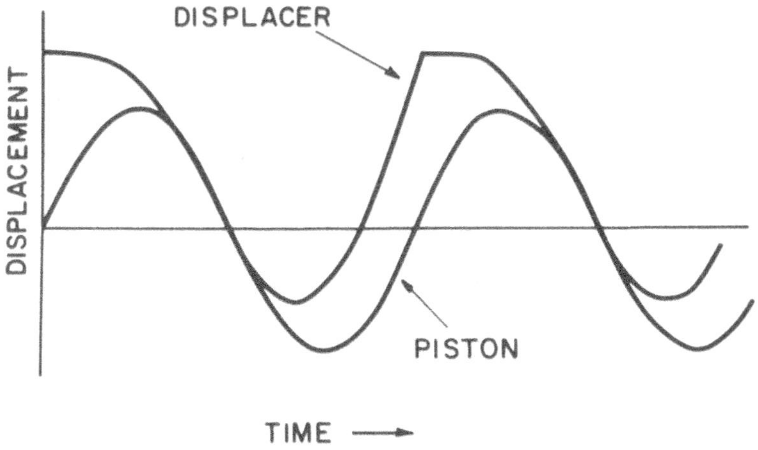

Figure 5.13 Piston and displacer motions in an overdriven single-cylinder hybrid.

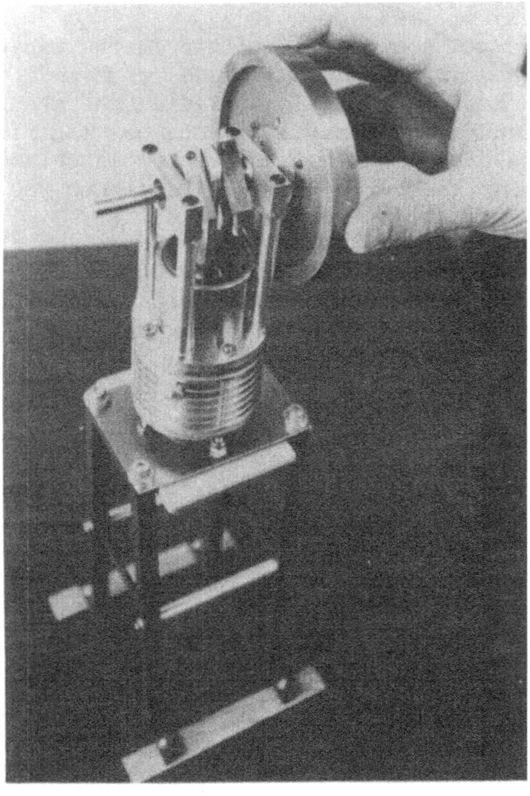

Figure 5.14 Single-cylinder Ringbom engine designed to operate in the overdriven mode.

together for a portion of the expansion stroke. This increases the specific power of the engine. The demonstration engine is shown in Figure 5.14. It has a swept volume of 14 cc and operates in the overdriven mode to a speed of just under 900 rpm on the heat of a simple bunsen burner. Further testing and development is underway, but the initial performance of this engine already points to renewed interest in the tandem Ringbom arrangement pioneered by Walker. It is apparently the simplest rotary shaft output Stirling engine of all and is, therefore, very attractive from the practical point of view.

NOMENCLATURE

A	cross-sectional area of displacer
A_P	cross-sectional area of piston
A_R	cross-sectional area of displacer rod
L_P	amplitude of piston motion = 1/2 piston stroke
L	maximum amplitude of displacer motion = 1/2 maximum displacer stroke
x_p	piston position at time t
x_d	displacer position at time t
V_e	instantaneous volume of expansion space
V_c	instantaneous volume of compression space
V_D	volume of dead space
T_E	temperature of working fluid in the expansion space
T_C	temperature of working fluid in the compression space
T_D	temperature of working fluid in the dead space
p	instantaneous pressure of the working fluid
p_m	cycle pressure at midstroke of piston and displacer
$\Delta p = p - p_m$	pressure difference across displacer rod
R	gas constant of working fluid
M	mass of working fluid
M_D	mass of displacer assembly
ω	angular velocity of crankshaft
K	$= [A_R M R T_C (1-\tau)]/[A M_D L^2 (1+\tau+\lambda\kappa+\sigma)^2]$
t_o	$= \omega^{-1} \sin^{-1} (\beta^{-1})$

Dimensionless Parameters

$\tau = T_C/T_E$

$\tau' = T_C/T_D$

$\lambda = L_P/L$

$\kappa = A_P/A$

$\sigma = \tau' V_D/AL$

$\gamma = \sqrt{K}/\omega$

$$\beta = \frac{\beta\kappa}{1-\tau}$$

$$\chi = \sqrt{\beta^2-1}$$

REFERENCES

Ringbom, O. (1907). Hot Air Engine. U.S. Patent No. 856102.

Senft, J.R. (1981a). The Hybrid Stirling Engine. Paper No. 819785, Proc. 16th I.E.C.E.C., Atlanta, Ga.

Senft, J.R. (1981b). Parameter Effects in Hybrid Stirling Engines. Consultant Report prepared for Univ. of Calgary, Dec.

Senft, J.R. (1981c). Proof of a Theorem on Hybrid Stirling Engines. UW-RF Research Report.

Senft, J.R. (1982a). A Hybrid Stirling Engine. Engineering in Miniature, Sept.-Nov.

Senft, J.R. (1982b). A Mathematical Model for Ringbom Engine Operation. UW-RF Research Report, Sept.

Senft, J.R. (1983). First Order Analysis of Ringbom Stirling Engines. Proc. 18th I.E.C.E.C., Paper No. 839125, Orlando, Florida, Aug.

Walker, G. (1973). Stirling Cycle Machines. Oxford University Press, Oxford.

Walker, G. (1980). Stirling Engines. Oxford University Press, Oxford.

Walker, G., Fauvel, R., Srinivasan, V., Gustafson, R. and van Benthem, J. (1982). Future Coal-Burning Stirling Engines. Proc. Stirling Engines Progress Towards Reality, pp. 1-9, I. Mech. Eng., London, (Univ. Reading, March).

Walker, G., Kentfield, J.A.C., Johnson, E., Fauvel, R. and Srinivasan, V. (1983). Coal-Fired Stirling Engines for Railway Locomotive and Stationary Power Applications. Inst. of Mech. Eng., London, June 6 Meeting.

CHAPTER 6 - DIAPHRAGM STIRLING ENGINES: THE HARWELL THERMO-MECHANICAL GENERATOR

E.H. Cooke-Yarborough studied Physics at Oxford University from 1937 to 1940, and then joined the Air Ministry Research Establishment (later re-named the Telecommunications Research Establishment), where he worked on airborne radar, on radar countermeasures and on guided weapons. He joined the Atomic Energy Research Establishment at Harwell in 1948, receiving Individual Merit promotion to Deputy Chief Scientist in 1955, and being appointed Head of the Electronics Division in 1957. In 1980 he was elected to the Fellowship of Engineering (whose membership consists of the 500 most prominent UK engineers) and was appointed Chief Research Scientist, Instrumentation and Applied Physics Division. He left Harwell to become a private Consultant in 1982.

He led the teams which designed and put into operation at Harwell the second digital computer to go into regular use in the UK and, later, what was the first all-transistor digital computer in the UK, and the second in the world. He also led the team which developed and put into service the Thermo-mechanical Generator. He has been jointly awarded four Premiums by the Institution of Electrical Engineers: - two for innovations in Nuclear Instrumentation, one for a paper on Opto-electronic Computer Logic, and one for the invention of the Stirling-cycle Thermo-mechanical Generator. He is the author of the book 'An Introduction to Transistor Circuits' and of many papers on Nuclear Instrumentation, Electronic Circuits and Energy Conversion. About 75 UK patents have been filed in his name.

G. Walker

INTRODUCTION

In free-piston Stirling engines, some of the more difficult design problems are concerned with wear and friction in the moving parts. Liquid lubricants cannot be used where they are exposed to the working gas, because they find their way into the hotter regions, where they decompose and clog the gas passages with solid matter. One approach is to use materials for the surfaces of piston and cylinder which will operate unlubricated with minimum wear. It has yet to be shown that this approach can lead to an engine operating life of more than a few thousand hours. Another approach is to use gas lubrication: some of the working gas being fed under pressure into the narrow space between piston and cylinder so as to keep the surfaces out of contact. This is said to be effective, but adds complication.

Another arrangement is to cantilever the piston on the end

of a piston rod which slides in guides outside the working volume. If there are parts of the piston surface which contact the cylinder wall, these will wear away, leaving the piston just out of contact. The problems of friction and wear are then transferred to the guides, which may be easier to lubricate.

Whichever method is used, the clearance between piston and cylinder has to be close (of the order of tens of micrometers) if blow-by losses are to be acceptably low.

Sealing between displacer and cylinder wall is less critical, since the pressure difference between the two ends of the displacer should be small, and some of the working gas can be allowed to pass through the clearance between displacer and cylinder. In the smaller engines this space is used as the only regenerator, and all the displaced gas passes through it.

The displacer does however present other problems. In a single cylinder free-piston engine the displacer is confined in a space bounded by the cylinder walls and cylinder head, and by the piston crown. It oscillates in this space with a phase in advance of that of the piston. Means must be provided for exciting the displacer into oscillation, and for guiding its motion. A method commonly used in free-piston engines is to mount the displacer on a rod passing through a cylindrical hole in the piston crown. This rod acts as a small-diameter piston which both guides the displacer and exerts on it an alternating force proportional to the alternating component of the pressure of the working gas. It can be shown that this force can cause the displacer to oscillate in the correct phase relative to the movement of the piston.

This rod passing through the piston crown is subject, on a smaller scale, to the same problems of sealing, wear and friction as the power piston itself.

With the engine at rest, there will be at least some static friction between the unlubricated sliding surfaces, which will prevent the free-piston engine from being reliably self-starting, since there must be at least some movement before any power can be developed to overcome friction.

In low-power engines (of the order of tens of watts) the effect of friction is particularly important. As the linear dimensions are scaled down, the energy produced per cycle decreases in proportion to the cube of these dimensions, whereas the area subject to dry friction decreases only with the square of these dimensions. The mechanical efficiency is therefore reduced.

Power at this sort of level is often required by installations at remote sites, which may be expected to operate for a year or more without attention. Propane-heated thermo-electric generators are often used, but have the disadvantage of low overall efficiency (about 3%), so a great deal of fuel has to be transported to the site.

A Stirling engine can offer much higher efficiency, but is only acceptable if it is as reliable as a semiconductor device. Thus a Stirling engine used in such an application should not wear significantly over a period of at least several years. 50,000 hours can be regarded as a minimum target. It is interesting to compare this with the life expected of an automobile engine. A car which has run for 100,000 miles at an average speed of 40 mph has run for only 2,500 hours. The operating life of a Stirling engine in these applications is therefore expected to be at least 20 times as long as that of an automobile engine, and without benefit of piston lubrication.

DIAPHRAGMS

The use of a metal diaphragm instead of a piston allows mechanical friction and wear to be eliminated (Cooke-Yarborough (1967)). It is, however, necessary to design the diaphragm carefully in order to ensure that these problems are not replaced by one of fatigue failure. It is well known that for most ferrous alloys there is a stress known as the "fatigue limit". If an alternating stress less than this limit is applied to the metal, it appears that this stress can be reversed an indefinite number of times without fatigue failure occurring. At higher stresses, eventual failure can be expected. The higher the stress, the earlier can failure be expected to occur, though there is a large variation in the fatigue lives of different samples of the same material subjected to the same stress. As the alternating stress is reduced towards the fatigue limit, the average number of stress cycles before failure increases very rapidly, and tends towards infinity at the fatigue limit.

To avoid extensive life testing of diaphragms and diaphragm materials, much use was made of published fatigue test results. The ferrous alloys for which most test results are available are the stainless steels. The different published fatigue tests were carried out by several different methods on a number of different compositions of stainless steels, prepared in different ways. Not surprisingly, the fatigue limits obtained varied from one set of tests to another over a range of about 2:1 (Cooke-Yarborough (1974a)). The results are summarized in Figure 6.1.

To make the best use of published results, it was decided to derive a "failure threshold stress" for each set of tests. This is defined as the mean of two stresses. One of these is the lowest stress at which any of the test specimens of a particular batch failed. (This will be just above the fatigue limit.) The other is the highest stress at which any of the test specimens of the same batch survived. Figure 6.1 shows that for any given material, these two stresses are very close together. The failure threshold stresses are indicated by circles. No failures occurred among specimens stressed at less than 95% of the failure threshold stress. This evidence suggests that if one sample survives when tested under a particular set of conditions, then, if the stress is reduced by 10%, all similar samples would be expected to survive under conditions

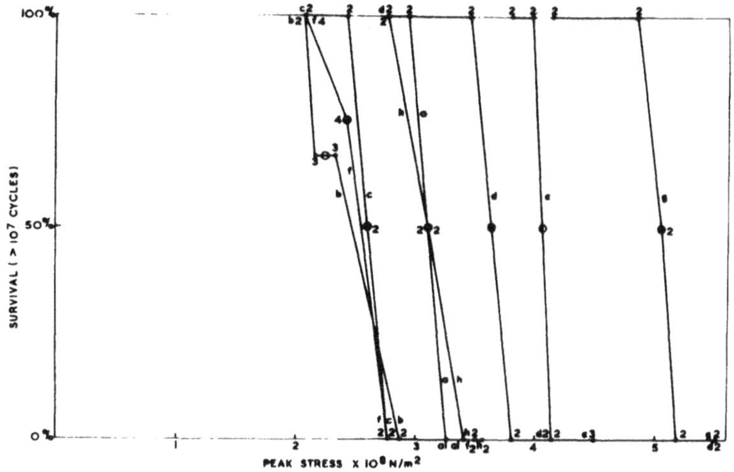

Figure 6.1 Typical relationship between the alternating stress S and the number of stress cycles N to fatigue failure.

which are otherwise similar. (This will not be true, of course, if any of the samples contains a flaw which raises the stress locally in a critical region.)

Most of the published fatigue test runs were ended after 10 or 100 million stress cycles. However, a Stirling engine running at 100 Hz would complete more than 10,000 million stress cycles in four years. The available information therefore has to be extrapolated by a very large factor. Since it is known that at stresses close to the fatigue limit, a very small reduction in stress results in a large increase in life before failure, it is not unreasonable to suppose that if the material is stressed to significantly below the fatigue limit, indefinite life will be obtained. An arbitrary decision was therefore made to limit the stress in the material to 2/3 of the failure threshold stress as defined above. The correctness of this decision is borne out by subsequent experience (Cooke-Yarborough (1982a)).

Having decided what alternating stress is acceptable, it is then necessary to ascertain whether a suitable diaphragm can be designed within these stress limits. At this point it is necessary to define one of the design objectives, namely the way in which the power produced by the diaphragm is to be coupled to the load. If the load is a liquid, then the diaphragm serves simply to separate the working gas from the liquid load. It transmits pressure, and does not need to have significant rigidity. (This is what is done in the small Stirling-cycle thermocompressors designed for powering artificial hearts.) If, however, the diaphragm is to drive a mechanism, such as an electrical alternator, it must apply an alternating force to a shaft. It must therefore be able to turn a pressure (distributed force) into a concentrated force, so must have a degree of rigidity. It is therefore attached centrally to a rigid hub which drives the output shaft, and at its outer edge to a rigid rim, fixed to the body of the engine.

The diaphragm is stressed when the hub deflects (Figure 6.2a). For a given hub deflection, the strain at the surface of the diaphragm is proportional to the thickness of the diaphragm material. It is also stressed by a pressure differential across it, as this causes the diaphragm to bulge between the hub and the outer rim (Figure 6.2b) by an amount which is inversely proportional to the cube

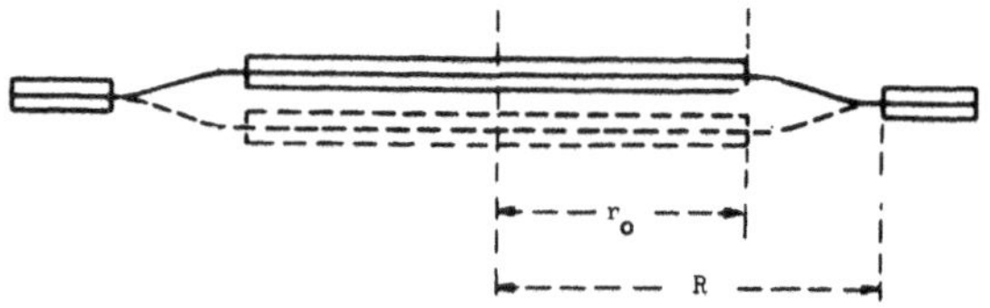

Figure 6.2 (a) Deflection of rigidly-mounted diaphragm. (b) Bulging of rigidly-mounted diaphragm under differential pressure. (Both exaggerated.)

of the diaphragm thickness. The radial curvature which this causes is, therefore, also inversely proportional to the cube of diaphragm thickness, and so the radial surface strain is inversely proportional to the square of the diaphragm thickness. In both cases the radial curvature, and the resultant radial surface stress, is at a maximum round the inner edge of the diaphragm, where it joins the hub.

For the present purpose it is a good enough approximation to assume that the gas pressure is near to its maximum at the point in the cycle when the volume of the working gas is at its minimum, which is when the diaphragm is at its maximum inward deflection. The radial surface stress near to the inner edge of the diaphragm is then the sum of the stress due to diaphragm deflection and that due to bulge. The stresses due to the two causes vary with diaphragm thickness in opposite directions, so, for a given maximum deflection, and a given maximum pressure differential, there will be an optimum thickness of diaphragm which minimizes the sum of the stresses. The deflection stress is then twice the bulge stress (Cooke-Yarborough (1968)).

Apart from contributing to diaphragm stress, bulging of the diaphragm has the effect of reducing the diaphragm swept volume, and it therefore reduces the engine power output. There is thus an advantage in making the diaphragm thickness rather greater than the optimum for minimum stress. If the diaphragm is made 26 percent thicker, the bulge volume is halved, yet the total stress is increased by only 5 percent.

There is also an optimum hub diameter, relative to the inner diameter of the rigid outer rim. If the hub is larger than this, the radial stress due to hub deflection will increase, while, if it is smaller, the bulge stress will increase. The ratio (c) between the outer and inner diameters which minimizes the total radial stress can be calculated to be about 1.8:1 (West (1969)).

The calculations summarized above led to the design of a flat diaphragm, stressed radially to a maximum of 200 megapascals (which is about 60 percent of the fatigue threshold stress for the 18/8 stainless steel used). Such a diaphragm was tested by subjecting it to alternating deflection at a frequency of 50 Hz in a test rig for two and a half years. There was no sign of deterioration at the end of this test. This diaphragm therefore survived 22,000 hours, during which it was subjected to 4,000 million stress cycles.

Although flat diaphragms were used in the early research engines, their stiffness is very sensitive to thermal expansion of the hub relative to the outer rim. This is important, since most of the heat rejected by the engine is deposited on the hub, which has to be provided with a means of cooling. The thermal sensitivity of the diaphragm is much reduced by corrugating the diaphragm circumferentially, as this allows differential thermal expansion to be accommodated with much less thermal stress.

The deflection and bulge stresses in corrugated diaphragms are difficult to calculate. However, approximate calculations (West (1971a)) show that quite shallow corrugations, having a depth of the order of twice the diaphragm thickness, will reduce, by a factor of 10, the sensitivity of the diaphragm stiffness to differential thermal expansion without increasing by more than a few percent the maximum radial bending stress near the inner edge.

To give an idea of the actual quantities involved in the design of a TMG, the text which follows will be illustrated with the quantities corresponding to a TMG giving a net electrical output of 25 watts. A diaphragm designed for such an engine, which operates at a mean pressure of one atmosphere, at a frequency of 110 Hz, has the following dimensions (West (1971b)):

Outer diameter	260 mm
Inner diameter	144 mm
Diaphragm thickness	0.9 mm
Stroke of hub	1.4 mm
Effective area	320 sq cm
Swept volume	45 cc

This diaphragm is illustrated in Fig. 6.3, together with the contoured cavity plate against which it works.

Figure 6.3 Corrugated diaphragm with rigidly-mounted hub. Contoured cold-cavity plate, with aperture in center for cylinder-displacer assembly, (after Cooke-Yarborough *et al* (1975)).

More recent work (Cooke-Yarborough (1976), Cooke-Yarborough (1982a)) has shown that there are important potential advantages in designing the diaphragm so that it is attached flexibly to the hub and to the outer rim. This can be done, for example, by clamping the inner and outer edges between rubber "O" rings. This relieves the edges of the diaphragm from the radial bending stresses described above, so that the diaphragm deflects in the shape of a shallow, truncated cone (Figure 6.4). The most important stress is now one which was neglected in the calculations for rigidly-mounted diaphragms: the circumferential surface stress caused by the conical deflection. Approximate calculations indicate that a diaphragm having the inner and outer diameters given above, and operating within the same stress limits, would perform as follows:

Diaphragm thickness	2.3 mm
Stroke of hub	4.2 mm
Effective area	320 cq sm
Swept volume	135 cc
Stiffness	335 N/mm

A diaphragm of this design can be deflected three times as far without

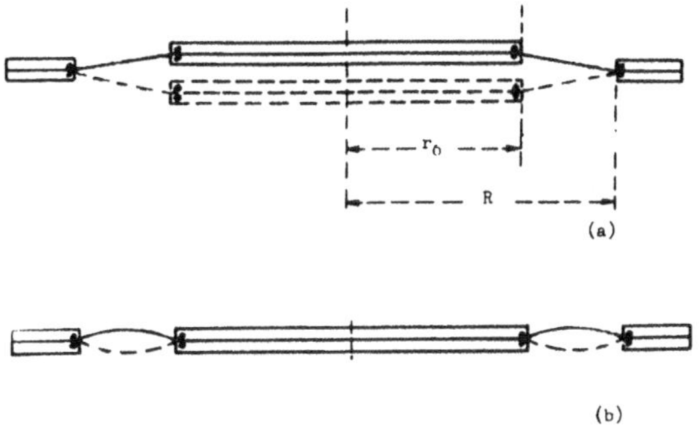

Figure 6.4 Comparison between deflection of rigidly-mounted diaphragm (a) and articulated diaphragm (b), (both exaggerated).

exceeding the stress limit, yet can be considerably thicker, and so operate at higher gas pressure differentials.

Diaphragms of this type have become known as Articulated Diaphragms. In addition to the above advantages, the flexible attachment to the hub makes the articulated diaphragm insensitive to differential thermal expansion, so corrugation is unnecessary. Because this is a relatively recent development, it has only recently been applied to diaphragm Stirling engines. It makes possible increases in both the swept volume and in the operating pressure, thus substantially increasing the power output obtainable from a diaphragm engine of given size.

From the dimensions given above, it is obvious that an engine using either type of power diaphragm is very different in shape and size from one using a power piston, because the diaphragm stroke is so much smaller than that of a piston providing the same swept volume, and the diaphragm diameter is very much larger.

DISPLACER DESIGN

In a TMG having a corrugated diaphragm, the diameter of the displacer is determined by the following practical considerations:

(a) The displacer spring is made smaller in overall diameter than the smallest corrugation of the diaphragm, since it cannot be contoured to match the corrugations.

(b) The cylinder is made smaller in diameter than the outer fixing points of the displacer spring, so as to provide space for a stationary surface to which the outer feet of the spring are bolted.

(c) There is a clearance between cylinder and displacer of

about 1 mm. This functions as the regenerator. To provide a direct path for the gas, the regenerator should have the same diameter as the outer pair of slots in the displacer spring.

Starting with the diaphragm dimensions given earlier, these considerations lead to a displacer diameter of about 130 mm. With a stroke of 2.5 mm, the volume swept by the displacer is 33 cc (West (1971b)).

Use of an articulated diaphragm without corrugations permits the diameter of the displacer spring to be increased, almost to the diaphragm diameter. This permits an increase in displacer diameter, giving a much enlarged swept volume, and increased power output (see section 'Operating Experience').

Although the area of the displacer is less than that of the diaphragm, it is, for a given power output, much larger than in the equivalent piston engine. Since the opposite ends of the displacer are at very different temperatures, it is important to ensure that there is not excessive heat leakage through this relatively large displacer area. To avoid the need for the displacer to be hermetically sealed, it is filled with the working gas (helium), and communicates with the working volume through a slow gas leak. Radiation and convection through the helium in the displacer body can be made acceptably small by use of thin metal baffles spot-welded to the inside walls of the displacer, but thermal conduction along the displacer walls (and along the walls of the cylinder) can contribute significantly to the thermal losses. For this reason these walls are made as thin as is practicable (0.4 mm in most of the engines built so far). Both displacer and cylinder are made by deep-drawing, the walls then being thinned-down by machining.

As already indicated, the displacer should be as light as possible. The cold end, where it is attached to the displacer spring, does however have to be quite strong and rigid, as substantial alternating bending moments are exerted on it by the spring.

In the 25-watt TMG, the displacer oscillates at 110 Hz, with a stroke of 2.5 mm, so its velocity at mid-stroke is 0.86 meters/sec. With a mass of 0.8 kg, its kinetic energy is 0.3 joules. This mechanical oscillation is damped by the viscous losses in the displaced gas. Calculation shows that the total viscous loss is of the order of 10 watts, or 15 millijoules per radian. Thus the displacer loses only about 5 percent of its mechanical energy per radian. It is therefore quite lightly damped (much less so than the diaphragm, which is loaded by the alternator). The natural frequency of the displacer oscillating on its spring is therefore the main factor which determines the frequency at which the engine runs.

To estimate the gas pressure change which would be brought about by displacer movement in an ideal, lossless engine, we make the simplifying assumption that all of the engine is either at the upper temperature T_e or the lower temperature T_c, and that there is a sharp dividing line between them half way along the displacer. When the displacer is at mid-stroke, the volume of gas which is at temperature T_e is V_h. When the diaphragm is at mid-stroke, the total volume of the working gas is V_m. If there is a phase difference of 90° between displacer and diaphragm movements, then when the diaphragm is at mid-stroke, the displacer is at one of the extremes of its stroke. The total volume swept by the displacer between these extremes is V_e.

Taking the mean working-gas pressure as P, an approximate calculation, based on expressions derived by Cooke-Yarborough (1974c) gives the pressure change resulting from movement of the displacer from mid-stroke to full stroke as:

$$\frac{P\,V_e}{2(V_m - V_h(T_e - T_c)/T_e)} \cdot \frac{T_e - T_c}{T_e} \qquad (6.1)$$

V_m includes all the various dead volumes, such as those in and around the displacer spring, some of which are difficult to measure directly. If the displacer motion is sinusoidal, then the pressure change resulting from this motion should also be sinusoidal.

To maximize the pressure change, V_m needs to be made as small as possible. A gas-volume measurement carried out on the 25-watt engine gave V_m = 180 cc. V_h is taken to be the mid-stroke volume of the heated cavity, plus half of the regenerator volume, and amounts to about 50 cc. Taking T_e as 723 K (445°C) and T_c as 313 K (40°C), and putting the working gas at atmospheric pressure (P = 100 kilopascals) yields a pressure change of 6.17 kilopascals at mid-stroke of the diaphragm, or about 6 percent of the pressure of the working gas.

If the diaphragm motion is sinusoidal, then, at mid-stroke, the rate of change of volume is $1/2\ V_o\omega$. The instantaneous power will be this rate of change, multiplied by the pressure produced by displacer motion (Equation (6.1)). If the displacer and diaphragm motions are in quadrature, then the peak of this occurs at mid-stroke of the diaphragm, and the mean power generated will be half of this peak power. If the displacer and diaphragm phases differ by θ, then the power generated is smaller by the factor $\sin\theta$.

Thus the power output of a lossless system is:

$$\frac{P\,\omega V_o\,V_e \sin\theta}{8(V_m - V_h(T_e - T_c)/T_e} \cdot \frac{T_e - T_c}{T_e}$$

Note that the second term represents the Carnot efficiency. To obtain the heat input to the engine, this term is made unity, while to obtain the heat rejected, it is made T_c/T_e.

In the 25-watt engine, with the diaphragm and displacer motions in quadrature, the above equation yields a power output of 48 watts rms.

The actual mechanical power output is found to be about two-thirds of this. The discrepancy can be attributed in part to viscous losses in the moving gas, and in part to imperfect thermal conduction between the working gas and the walls of the working spaces, so that the temperatures of the gas deviate from the wall temperatures while the gas is being expanded or compressed. These losses are discussed in the section 'Increasing the Generator Power'.

THE ALTERNATOR

The alternator is required to convert the mechanical oscillation delivered by the diaphragm into voltage and current applied to an electrical load.

The diaphragm hub oscillates with a short stroke, so, for a given power output, the alternating forces are correspondingly large. If the oscillation is sinusoidal, then at mid-stroke the peak mechanical power output is equal to the product of the diaphragm hub force and the diaphragm hub velocity. (The mean power output is then one half of this.) This force must be balanced by the force generated in the alternator by the instantaneous electric current flowing through its windings.

A moving-coil alternator is not very suitable when the mechanical stroke is small (Cooke-Yarborough (1967)). The reasons are as follows. At a given power level, if the stroke is reduced, the force exerted by the diaphragm must be increased, and a given current flowing in the load must balance this increased force. The number of turns in the coil therefore has to be increased. The coil has, however, to be accommodated within the volume of the magnetic gap, so the gauge of the wire must be reduced. The result is that the coil resistance is inversely proportional to the square of the stroke, and so is the power lost in the coil. This can be offset by increasing the coil volume, and with it the volume of the magnetic gap, but a correspondingly more powerful field magnet is then required to maintain the same flux density in the gap.

If a small mechanical movement is used to vary an air gap in a magnetic circuit, a very large change of magnetic flux can result, because of the enormous difference between the permeability of air and that of magnetic iron. After study of a number of alternatives (Franklin (1970a)), a balanced variable-gap flux-switching configuration was chosen (Figure 6.5). A moving magnet is used, and movements of less than one millimeter in either direction can switch the flux almost completely from one coil to the other. The coils themselves, being outside the magnetic gaps, can be large and of low resistance. The magnetic armature tends to be attracted to

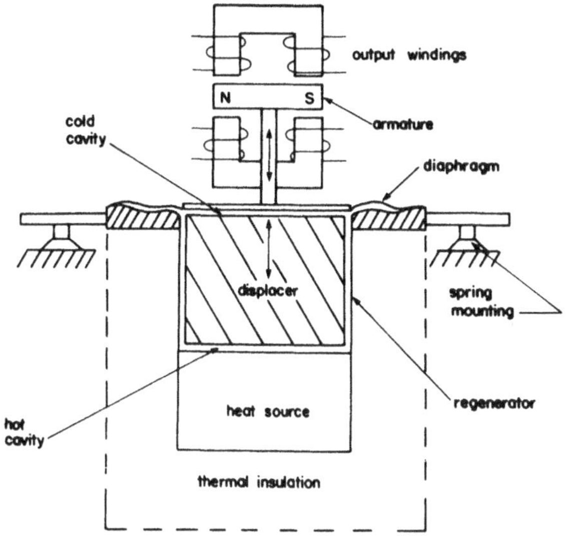

Figure 6.5 General arrangement of a thermo-mechanical generator, (after Cooke-Yarborough *et al* (1974)).

either pair of pole pieces. This is equivalent to a negative stiffness so, to stabilize the armature about its mid-position, the positive stiffness provided by the diaphragm has to be augmented by strong centering springs.

As in the case of the engine, it is useful to consider the behaviour of the alternator at mid-stroke. Here the forces due to the permanent-magnet flux in the opposing gaps are balanced, and the magnetic force on the armature is due to the flux produced the current i flowing in the windings. If the reluctance of the iron is small compared with that of the air gaps, it can be shown that this force is:

$$\frac{B\,A_m\,N\,i}{2X} \qquad (6.2)$$

where
- B is the flux density in the permanent magnet,
- A_m is the cross-sectional area of the permanent magnet,
- N is the total number of turns in the two coils,
- i is the instantaneous coil current,
- X is the width of each air gap at mid-stroke.

The alternator used in the 25-watt system has the following parameters:

B = 0.7 tesla
A_m = 13.7 sq cm

N = 320 turns
X = 1 mm

From Equation (6.2), the force which is exerted on the armature in order to produce a load current is therefore 153 Newtons per ampere.

The effective area of the diaphragm is A_o, so from Equation (6.1) the force produced at mid-stroke of the diaphragm by gas displacement is:

$$\frac{P\ V_e\ A_o}{2(V_m - V_h(T_e - T_c)/T_e)} \quad \frac{T_e - T_c}{T_e} \tag{6.3}$$

Combining Equations (6.2) and (6.3) yields the following expression for the peak load current:

$$i = \frac{2X}{B\ A_m\ N} \cdot \frac{P\ V_e\ A_o}{2(V_m - V_h(T_e - T_c)/T_e)} \cdot \frac{T_e - T_c}{T_e} \tag{6.4}$$

The effective area (A_o) of the diaphragm of the 25-watt engine is 320 sq cm, so the relationship between pressure and load current can be calculated from Equation (6.2) to be 4.8 kilopascals per ampere. The pressure at mid-stroke produced by displacer movement has been calculated from Equation (6.1) above to be 6.17 kilopascals. The force exerted by the diaphragm on the alternator at mid-stroke is, therefore, 198 Newtons, giving a peak load current (i) of 1.3 amps (0.91 amps rms).

The emf generated in the alternator is proportional to armature velocity, and at mid-stroke can be shown to be:

$$\frac{B\ A_m\ N\ y_o\ \omega}{4X} \tag{6.5}$$

where y_o is the stroke of the diaphragm hub and the armature, and ω is the angular operating frequency in radians per second. In the case of the 25-watt engine with the alternator parameters given earlier, running at 110 Hz, with a stroke of 1.4 mm, the peak emf is 74.2 volts (52.5 v rms).

The alternator has considerable series inductance, which severely limits the current which the generated emf can cause to flow in a resistive load. This can be largely overcome by tuning out the inductance with a series capacitor which resonates with the inductance at the operating frequency, so as to produce a nearly-zero series impedance at this frequency. The alternator used with the 25-watt engine has an inductance of 47 millihenries, so the capacitance value required is 50 microfarads. In what follows, it is assumed that the inductance has been tuned out in this way, so that the voltage appearing across the load is equal to the generated emf.

The resistance of the electrical load is obtained by dividing the load voltage (Equation (6.5)) by the load current (Equation (6.4)). When the displacer swept volume V_e is replaced by the product of the displacer stroke x and the displacer area A_d, this resistance is given by:

$$\frac{(B\ A_m\ N)^2}{4X^2} \cdot \frac{\omega\ y_o}{x} \cdot \frac{(V_m - V_h(T_e - T_c)/T_e)}{P\ A_d\ A_o} \cdot \frac{T_e}{(T_e - T_c)} \qquad (6.6)$$

This equation contains y_o/x, the ratio of diaphragm stroke to displacer stroke. The factors which determine this ratio are considered in the section 'Losses'. In the case of the lossless 25-watt engine, the load resistance given by the above equation is 57.4 ohms.

To maintain oscillation at constant amplitude in such an engine, the resistance of a linear load must be just equal to the value given by the above equation. A lower resistance will cause the oscillation to die down, while a higher resistance would, in a purely linear system, cause the oscillation to build up indefinitely. In practice the relationship between oscillation amplitude and output current is not linear and tends to saturate at high displacer amplitudes. This is partly due to the imperfect heat transfer mentioned earlier, and partly to the narrowing of the cavities at the peaks of the strokes causing increased gas flow loss at these parts of the cycle. Consequently, as the load resistance is raised progressively, the output current rises less rapidly than the output voltage, so the generated power rises progressively with load resistance, and reasonably stable operation is possible with a resistive load.

A constant-voltage load is much to be preferred, as this tends to keep y_o constant. The oscillation amplitude can then be closely controlled at a level just short of that which would cause the displacer and diaphragm to reach the mechanical limits of their respective strokes, thus obtaining the largest power output of which a given engine is capable. With the load voltage fixed, the oscillation amplitude can if necessary be adjusted over a limited range by adjusting the flux density B of the permanent magnet (Equation (6.5)).

In many applications, the TMG is connected via a full-wave rectifier to charge a battery. This serves as an approximately constant-voltage load, and so acts to stabilize the oscillation amplitude of the TMG. The voltage appearing at the input to the full-wave rectifier is almost a square wave, but because of the presence of the series-resonant circuit, which has a high impedance at harmonics of the operating frequency, the alternator current and the generated emf remain nearly sinusoidal, and the battery charging current consists of a series of half-sinusoids which, in a lossless system, would have the peak value given by Equation (6.4).

SYSTEM DYNAMICS

No mention has yet been made of the way in which the displacer is maintained in oscillation. This is done through vibration of the body of the engine, which is spring-mounted. If the diaphragm hub, the alternator armature and the other components which oscillate with them, have a total mass M_o, and the mass of the engine body is M, then, to conserve momentum, the engine body will vibrate with a stroke which is a fraction $M_o/(M + M_o)$ of the diaphragm stroke y_o (which is measured relative to the engine body), and in the opposite direction.

This body vibration exerts an alternating force on the displacer through the stiffness S_d of the displacer spring. If this vibration is at the displacer resonant frequency, then the phase of the displacer oscillation will be such that the component of displacer-spring force due to body vibration is at a maximum when the displacer is at mid-stroke. This force is then:

$$\pm \frac{M_o\, y_o\, S_d}{2(M+M_o)} \qquad (6.7)$$

The peak power fed to the displacer is then the product of this force and the displacer velocity at mid-stroke. The mean power is one half of this, and in equilibrium, is equal to the power P_d used by the displacer in overcoming the viscous loss in displacing the gas, and in maintaining a displacer stroke x. Thus:

$$P_d = \frac{M_o\, x\, y_o\, S_d\, \omega}{8(M+M_o)} \qquad (6.8)$$

The pressure changes brought about by displacing the gas have already been shown to be in phase with the displacer movement which causes them. To obtain the desired 90° phase relationship between these pressure changes and the resultant diaphragm movement, the mass M_o associated with the diaphragm hub should resonate, at the operating frequency, with the associated stiffness S_o. If M_o = M, then:

$$S_o = M_o\, \omega^2$$

This stiffness is made up of a number of components: the compression stiffness of the working gas, the mechanical stiffness of the diaphragm, the negative stiffness due to the magnetic forces in the alternator, and the positive stiffness of the alternator springs required to overcome this negative stiffness. In the 25-watt TMG, S_o = 600 Newtons/mm approximately.

From Equation (6.8), the engine body mass M required for a displacer stroke x is:

$$M = \frac{x\, y_o\, S_d\, S_o}{8P_d\,\omega} \qquad (6.9)$$

The mechanical resonance of the displacer mass on its spring determines the operating frequency. To maximize power output, this frequency should be as high as possible, so the displacer-spring stiffnes S_d is made as great as consistent with acceptable stress. In the 25-watt TMG, S_d is 400 N/mm and P_d is calculated to be 10 watts. The body mass of the machine should, therefore, be 32 kg. This is much smaller than the actual body mass of this machine, which is about 100 kg.

For this reason, the first designs were provided with magnetic displacer drive (a permanent magnet in the diaphragm attracting a laminated armature in the end of the displacer). In the event, this was found to be unnecessary, even with the radioisotope heated version (Cooke-Yarborough *et al* (1975)), which has a shielded radioisotope heat source weighing 250 kg included in the body mass.

The reason for this unexpectedly large displacer drive is not well understood, but is believed to be associated with flexing of the TMG body near the points of attachment of the displacer spring. Later designs, of more rigid construction, show behaviour much closer to that predicted by Equation (6.9).

The total power generated is the product of current and voltage (Equations (6.4) and (6.5)) and is proportional to the product of the displacer and diaphragm swept volumes V_e and V_o. It is, therefore, also proportional to the product of their strokes x and y_o. Thus Equation (6.8) shows that in a linear system a constant fraction of the generated power is diverted to excite the displacer.

At low amplitudes the system is almost entirely linear, and static friction is entirely absent, so, provided that the load resistance exceeds the value given by Equation (6.6) by enough to make good the viscous losses, the engine self-starts reliably from rest. It can be stopped and re-started by applying and removing an electrical short-circuit across the load.

There are two tuning adjustments which are needed when an engine is first commissioned. First, the mass M_o associated with the diaphragm hub and armature has to resonate with the associated stiffnesses S_o at the operating frequency determined by the displacer. This is not critical, because of the heavy damping provided by the load, and it has not been found necessary to vary the calculated mass. (Minor adjustment to the stiffness is done by controlling the pressure of the working gas.) Second, the effective mass M of the engine body has to be adjusted to give the correct ratio between diaphragm and displacer oscillation amplitudes. Quite small additions to the engine mass have been found to give the correct ratio. On some engines a resonant mass on a spring attached to the engine along its center line has been used for this adjustment. This can be tuned to either side of resonance so as either to add or to subtract an effective mass which can be up to ten times the actual mass on the spring.

In some applications, a substantial reduction in engine body mass may be desirable, and the displacer will then be over-excited. The displacer will also be over-excited if the diaphragm stroke is increased, e.g. as a result of using an articulated diaphragm. The excitation of the displacer can be reduced by including a direct elastic coupling between diaphragm hub and displacer, as this exerts an alternating force on the displacer which is in antiphase with that exerted through the displacer spring by vibration of the engine body. If the stiffness of this coupling is S_f, then, from Equation (6.9) it can be shown that the engine body mass can be reduced to:

$$M = \frac{x\, y_o\, S_d\, S_o}{4\omega(2P_d + x\, y_o\, S_f\, \omega)} \qquad (6.10)$$

LOSSES

Most of the above discussion has been simplified by neglecting losses. However, it has already been indicated that these losses cause the efficiency to fall well below that of a lossless Stirling engine, which would have the full efficiency of the Carnot cycle. The essence of all Stirling engine design is to seek compromises which make the total of all the losses as small as possible.

The losses fall into the following main categories:

1) Heat leakage from the heated end to the cooled end, or to the environment.
2) Incomplete heating and cooling of the working gas.
3) Gas friction loss.
4) Electrical and magnetic losses in the alternator.

Mechanical friction is also important in most Stirling engines, but plays no signficant part in the TMG.

Loss computations which attempt to take all or even most significant factors into account tend to be so complex as to obscure real understanding of the processes involved. Moreover, some losses are very sensitive to certain engine dimensions, and these dimensions may in practice neither be uniform, nor indeed accurately known, so that the potential accuracy of such computations is not in practice realised. Consequently, an approach which is much more approximate may give a qualitative understanding of at least the direction in which the losses will vary with engine operating parameters, and therefore guide the experimenter in varying these parameters to improve performance.

In the TMG, all gas passages and cavities are shallow. This leads to several simplifications. First, in calculating the mechanical power loss due to the movement of the working gas, all gas flows can be regarded as laminar (West *et al* (1968)), and the kinetic component can be neglected (West (1970)). Consequently, the power absorbed by the flow of gas along the gap between two parallel surfaces is proportional to the square of the gas velocity, and inversely proportional to the cube of the gap between the surfaces (Franklin (1970)).

A second simplification is that if the heat transfer is rapid enough, the temperature difference between the gas and the surfaces between which it lies, can be regarded as an instantaneous response to its rate of compression or expansion or to the rate of change of surface temperature as it flows along a temperature gradient (e.g. through the regenerator). This temperature difference is then proportional to the square of the spacing between the surfaces. In a TMG running at 110 Hz with helium at one atmosphere pressure, this simplification is valid only for gaps up to about 1 mm (Ryden (1976)).

A nearly instantaneous gas-temperature response leads to a further simplification if also the displacer movement is exactly in quadrature with diaphragm movement (which in a lossless system maximizes power output), for then many of the losses resulting from the two movements can be regarded as independent, and described by separate expressions.

Viscous Losses

The movements of displacer and diaphragm can be taken as being sinusoidal (Howlett (1970)) and the assumption that the resultant gas pressure changes are also sinusoidal leads to only slight errors in calculating the lossless power output (Cooke-Yarborough (1974)). Likewise, the gas flow can be taken as being sinusoidal, so the viscous loss in the gas flowing along a constant gap between stationary surfaces follows a sine-squared law (analogous to the power dissipated in a resistance by the flow of an alternating current). Thus calculation of the power dissipated at one known part of the cycle allows the average power dissipated to be calculated. In particular, the peak power dissipated is just twice the average power.

The sine-squared loss approximation is not valid in the case of gas flowing into or out of the heated or cooled cavities, since here the gap is varying with time, and most of the power is dissipated near to the brief period when the two surfaces are closest together, as was shown experimentally at an early stage in the development of the TMG (Cooke-Yarborough *et al* (1968)). To compute this loss, it is necessary to integrate throughout the cycle (West *et al* (1968)). In the cooled cavity there are three such gaps: that between the diaphragm hub and the splitter plate at the center of the displacer spring; that between the splitter plate and the cooled end of the displacer; and that between the diaphragm and the stationary surface which faces it. Each of these gaps varies with a different amplitude and phase. A more precise computation would take into account also the varying gaps associated with the displacer spring.

Regenerator Losses

As with other Stirling engines, the behaviour of the gas in the regenerator is quite complex, since the gas pressure is varying during the cycle and so, therefore, is the mass of gas in the regenerator. As a result, the gas flow rates at the two ends of the regenerator are generally unequal, and may be in opposite directions at certain points of the cycle (Franklin (1968)). Neglecting for the moment these complications, it will be seen from what has been said earlier that the viscous gas flow loss is inversely proportional to the cube of the gap between displacer and cylinder wall. It can also be shown (Franklin (1970b)) that the thermal loss through the regenerator is directly proportional to this gap. There is thus an optimum gap. In choosing this optimum it has to be borne in mind that the viscous loss is from mechanical power already converted from heat at the engine efficiency, whereas the thermal loss is from the unconverted heat. In an optimized design, therefore, the viscous loss will be much smaller than the thermal loss.

There are advantages in subdividing the regenerator by spacing one or more thin-walled cylinders between the displacer and the cylinder wall proper. Not only does this permit a more favourable compromise between viscous and thermal losses, but it also relaxes the tolerances required of the positioning of the displacer relative to the innermost of these cylinders (Cooke-Yarborough (1982b)).

Imperfect Heat Conduction

While the diaphragm is moving inwards, the gas throughout the working space is being compressed, so there is a flow of heat from the gas to the walls throughout the engine. Because the thermal conductivity of the gas is finite, this will everywhere cause a local temperature rise which is proportional to the amount of gas between the surfaces (which is proportional to the gap) and to the average distance in the gas through which this heat must flow (again proportional to the gap). The local temperature rise is therefore proportional to the square of the gap in each region (Cooke-Yarborough *et al* (1968), Appendix 2, Ryden (1976)). Each of these local temperature rises will contribute to an overall pressure rise in proportion to the amount of gas in that region. This pressure rise opposes the inward movement of the diaphragm, and therefore represents a mechanical loss. The same happens, in reverse, when the diaphragm moves outwards. Imperfect thermal conduction anywhere in the working volume can therefore contribute to this loss.

The volume of the heated cavity is largest while expansion is taking place, so here the temperature drop during expansion is greater than the rise during compression, and this lowers the mean temperature in this cavity. The mean temperature of the cooled cavity rises for the same reason. This, therefore, reduces the effective temperature difference between which the engine is working, and, therefore, reduces the thermal efficiency.

Computed Losses

Many loss computations have been performed, with varying degrees of approximation. Table 6.1 gives one set of results obtained for the 25-watt (nominal) TMG running at 450°C, computed by Howlett (1970).

Table 6.1
Computed Losses in the 25-Watt TMG

	Watts
Lossless Power	50.2
Gas Flow Losses	
In Heated Cavity	0.11
In Regenerator	5.57
In Cooled Cavity	4.05
Heat Transfer Losses in Cavities (Total)	6.1
Net Mechanical Power Output	34.37
Electrical Power Output (Alternator Efficiency 90 Percent)	30.93

Table 6.1 indicates the order of magnitude of the quantities involved, though it does not include the heat-transfer losses in the non-varying regions of the working volume (Ryden (1976)), so probably under-estimates the total losses. On the other hand, the alternator losses are probably over-estimated, even though the windings of the alternator used have about four times the resistance of the design described by Franklin (1970a).

The above results do not include direct heat leakage from the heated end, or heat leakage through the regenerator. The former is calculated by Franklin (1971) and the latter by West (1971b). However, actual heat loss through the regenerator has always exceeded the computed value by a substantial amount, and the model used clearly needs to be improved in this respect.

INCREASING THE GENERATOR POWER

The TMG was originally developed to make efficient use of the heat produced by radioisotopes, particularly Strontium-90. The power level of 25 watts was chosen because it was believed that a radioisotope source, with shielding, for any larger generator would be unacceptably expensive. In the event, it was found that there was a greater need for a version heated by propane, and so only one of the early generators used a radioisotope heat source (Cooke-Yarborough *et al* (1975)).

The removal of the limitation of the cost and power of the heat source led to the demand for higher-powered versions of the TMG. In one particular case, an output of at least 60 watts ac was required for powering a major lighthouse. The requirement was to obtain this power from an existing 25-watt TMG with the fewest possible modifications.

The only ways to increase power output without mechanical modifications are to increase the operating temperature T_e of the heated end, and to increase the pressure P of the working gas. Neither of these affects the generated emf (Equation (6.5)), but both increase the load current (Equation (6.4)).

The 25-watt TMG operates with the working gas near to atmospheric pressure, and early versions had the outer side of the diaphragm exposed to the air. In order to eliminate the effect of changes in air pressure on the mean position of the diaphragm, and to provide a helium reservoir, later versions had the enclosure which contains the alternator and the outer side of the diaphragm sealed from the atmosphere and filled with helium at the working pressure. This made it possible to raise the helium pressure without imposing a steady pressure differential on the diaphragm.

The increased gas pressure does, however, increase the alternating bulge stress on the diaphragm, but it was calculated (Cooke-Yarborough (1977a)) that by slightly reducing the diaphragm stroke y_o (from 1.4 mm to 1.1 mm), the total stress can be kept within the limits already established in the section 'Diaphragms'. This reduces the generated emf (Equation (6.5)) from the 52.5 v calculated in the Section 'Displacer Design' to 41.25 v. This emf was further reduced, by reducing the permanent-magnet flux density B from 0.7 Tesla to 0.53 Tesla, in order to provide the reduced output voltage of 31.2 v required in this application. Each of these changes has the effect of increasing the alternator output current, and as a result the copper loss in the alternator is calculated to have risen to 3 watts.

The heat loss through the regenerator is expected to increase with the square of the working-gas pressure (Franklin (1970)), and is in any case larger than calculated. Experiments with increased working-gas pressure showed that little was to be gained by going above 2.25 atmospheres. This is believed to be due to the sixfold increase in heat lost through the regnerator, and to the resultant heavy thermal loading on the heated and cooled cavities, causing a reduction in the effective temperature difference between them.

The hot-end temperature was raised to 600°C, the maximum, it was thought, for the braze used, and an ac electrical power output of 60 watts was just achieved. This proved adequate for the lighthouse, even in mid-winter (Cooke-Yarborough *et al* (1980a)).

This design was by no means optimized for this power level, and the efficiency is slightly less than that of the 25-watt engines.

Because of the large heat loss through the regenerator, a three-fold increase in radiator area is only just enough to cool the engine adequately.

Means of obtaining substantial further increases in power output have been described (Cooke-Yarborough (1982a)). The diameter of the displacer is doubled, and this, with the displacer stroke left unchanged, quadruples the displacer swept volume V_e. The regeneration is improved by using coaxial splitters in the regenerator space (Cooke-Yarborough (1982b)). The diaphragm stroke and swept volume are increased by using an articulated diaphragm.

It is desirable for a number of reasons to keep the overall dimensions of the TMG unchanged. Thus the outer diameter of the diaphragm cannot be doubled, as in the case of the displacer. However, the original corrugated diaphragm did not make best use of the available space, and the articulated diaphragm does not need a heavy clamping ring, so the outer radius of the articulated diaphragm can be increased from the original 130 mm up to 176 mm without increasing the overall diameter of the TMG. The articulated diaphragm is less critical than a rigidly-mounted corrugated diaphragm in respect of the ratio between the outer and inner radii, so the radius of the rigid hub of the articulated diaphragm can safely be doubled, to equal the new radius of the displacer.

The effect of these two increases is to increase the effective area of the diaphragm by a factor of 2.25. Because articulation reduces the bending stresses, the stroke can be restored to that of the 25-watt design (1.4 mm), so the swept volume is increased by a factor of 2.8, giving a corresponding increase in power output over that of the 60-watt engine, to 168 watts.

Increasing the stroke by a further factor of 1.4 (to 2 mm) would provide a four-fold increase in swept volume, with a corresponding further increase in power output over that of the 60-watt TMG. The maximum diaphragm stress is calculated still to be less than half of that in the corrugated diaphragm of the 60-watt TMG, so still further increases in power output may be achievable by further increases in stroke.

OPERATING EXPERIENCE

The basic design of the TMG was arrived at as a result of a research programme which began in 1965 (Cooke-Yarborough *et al* (1974)) and established the basic dimensions (West (1971)), which were closely followed in later designs.

The development programme which followed aimed at a design which could be manufactured. Three development TMGs were built (the D-series), each representing progress towards this objective. Each incorporated a different source of heat; the first was propane-heated, the second was heated by radioisotope while the third was electrically heated.

This was followed by a field-trial programme, for which four propane-heated field-trial generators were built (the F-series), two for powering the UK National Data Buoy, and two for powering marine lights.

The main characteristics of all these generators are set out in Table 6.2. The propane-heated D1 machine ran for several thousand hours in the laboratory. Although subject to flue losses and other thermal losses comparable to those encountered in propane-heated thermo-electric generators, it nevertheless achieved an overall efficiency about three times better. It was later used for demonstration, and ran at the British Genius Exhibition in London throughout the Summer of 1977.

Table 6.2
Performance of Thermo-Mechanical Generators

Application	Power Watts AC	Power Watts DC	Propane Used kg/Year	Efficiency to AC	Efficiency to DC	Operating Hours to April 1983
			Development (D) Machines			
D1 Lab	31.7	-	196	10.2	-	9,000
D2 Lab	10.7 (initial)	-	Isotope	7.7	-	72,000
D3 Lab	37.5	32.3	Electrical	16.9	14.5	small
			Field-Trial (F) Machines			
Data Buoy:						
F1	24.5	18.9	190	8.1	6.25	10,000
F2	-	27	166	-	10.7	21,000
AGA						
F3 (original)	-	18.5	-	-	-	small
Lighthouse (up-rated)	65	58	450	9.1	8.1	23,000
F5 (Trinity House)	31	24.7	164	12	9.5	small
				Total Hours:		135,000

The D2 TMG (Cooke-Yarborough *et al* (1975)) is illustrated in Figure 6.6. It is heated by strontium-90 titanate in a double-walled, welded capsule, inside a heavy-alloy radiation shield weighing 250 kg. The whole shield is thermally insulated and heated by the radioisotope source. It mounts directly onto the hot end of the cylinder of the TMG, which in this installation is inverted, with the diaphragm and alternator below the cylinder. The shield is designed to take a radioisotope heat source delivering up to 250 watts (thermal), which would enable the TMG to provide an electrical power output of 25 watts, with an overall efficiency of 10 percent. However,

Figure 6.6 Radioisotope powered TMG, (after Cooke-Yarborough *et al* (1975)).

a source readily available in 1974 was one delivering only 138 watts, so nearly half of the limited heat available from this source was lost through thermal conduction. As a result the power output was only 10.7 watts, and the overall efficiency 7.75 percent.

The power output of all radioisotope generators falls off with time as the source decays. Table 6.3 compares the performance of the system in November 1974 with its performance in April 1983. This shows that the fall in output can be attributed mainly to the reduction in the amount of heat available from the radioisotope source, and that the efficiency of the TMG changed very little over the period of eight and a half years. Although the overall efficiency obtained initially was less than the 10 percent which would have been obtained with a larger source, it is nevertheless far superior to what would be obtained from a similar source by thermo-electric conversion.

Because some loss of helium could be expected by diffusion through the various O-rings used in assembly, an external reservoir of 7 liters is connected through a slow leak to the working volume. Up to the beginning of 1979, the helium was replenished three times, and for this purpose the machine had to be briefly stopped. Thereafter,

Table 6.3
Performance of Radioisotope Heated TMG

	November 1976	April 1983
Output of heat source	138 w	112 w*
Temperature °C	292	326
Thermal conduction loss	60 w	66.3 w
Net heat to TMG	78 w	45.6 w
Electrical power output	10.7 w	5.2 w
TMG efficiency	13.7 %	11.3 %
Overall efficiency	7.75 %	4.6 %

*Calculated from isotope decay.

it ran for more than four years without requiring further replenishment. The field-trial generators use the cover over the alternator as a helium reservoir. This has a capacity of nearly 40 liters, and loss of helium by diffusion should be at the rate of less than 1 percent per year.

The electrically-heated D3 TMG incorporates many of the lessons learned from the other machines, and gave the results shown in Table 6.2. These show the advantage to be gained by raising the operating temperature. It has a sealed alternator cover, and resembles the field-trial generators in most respects.

The first field trial of a TMG was in the National Data Buoy. This is an unmanned, moored, wave-riding discus buoy 7.5 meters in diameter, equipped with a great variety of meteorological and oceanographic instrumentation, on-board data processing and a high-frequency radio transmitter. A major part of the electrical load is due to this transmitter, which is required to transmit real-time wave data continuously for periods of 20 minutes every three hours, in addition to shorter hourly data transmissions. The Buoy is equipped with two 100-gallon propane tanks, so as to provide the TMG with enough fuel for a two-year mission at sea without refuelling.

The F2 TMG was installed in the Summer of 1975, and experience in the North Sea up to the end of 1976 is recorded by Cooke-Yarborough *et al* (1977b). For 15 percent of the time propane was not available, because of a total loss through a leak in the buoy pipework. There were some problems of propane contamination, and buoy motion at sea showed up a weak joint in the pipework carrying pentane coolant to the radiator. The burner never quite reached the design temperature. Nevertheless, the TMG provided power for 90% of the time that propane was available.

By early 1977, the propane again became exhausted, possibly due to leakage resulting from deterioration of non-metallic pipes on the exposed tank tops. The F2 TMG was returned to the laboratory in September 1977. The operating experience with this generator, and subsequent examination in the laboratory (Cooke-Yarborough (1980b)) enabled a number of improvements to be incorporated in the F1 generator before it was installed in November 1977 during the refit of the Data Buoy.

The F1 machine was run continuously from March 1978. The Buoy was on station (165 miles south-west of Land's End) by the end of July 1978.

This TMG met the Buoy's power requirements with a substantial margin, but the burner temperature slowly fell during the following year, requiring the propane supply pressure to be increased. Contamination of the helium was suspected, so in the summer of 1979 this was replaced at sea. Performance immediately returned to normal, which confirmed that contamination was the cause, probably due to entry of air or coolant by leakage, not diffusion. Efficiency again fell slowly, and by January 1980 the propane was exhausted. The TMG had powered the Buoy for 21 months continuously, except for the brief stop to replace the helium. The Buoy remained at sea for another two years, being powered by primary batteries carried to the Buoy twice a year. Had it been possible instead to carry cylinders of propane to the Buoy to refuel the TMG, the weight transferred in these difficult operations could have been reduced to a quarter of the weight of the batteries.

In the summer of 1982, this TMG was recovered from the Data Buoy and taken to the laboratory of HoMach Systems Ltd., where it started up immediately on being supplied with propane. In 1983 it was put into service as a test-bed for electronic control systems, still with the same charge of helium which it was given at sea in 1979.

The F3 and F5 TMGs were purchased by AGA Navigation Aids Ltd. The F5 machine was supplied by AGA to the Trinity House Lighthouse Service in 1976, for evaluation at their Dungeness station. Apart from the few modifications common to all machines, the only incident was a partial jet obstruction. The machine was not, however, run for any length of time, though it was taken by sea to China and back. When it was re-tested in the AGA laboratories in the summer of 1979, the performance was found to be identical to that measured three years earlier, showing that helium loss by diffusion is negligible. In 1982 it was put into service by HoMach Systems Ltd. as a demonstration machine, still giving its original power output, and still with its original charge of helium.

The F3 machine was used by AGA for demonstration, and at the IALA Conference at Ottawa in the summer of 1975 it was run continuously for two weeks from a small propane cylinder in the ballroom of the conference hotel. The exhaust of these machines is so clean, and the oxygen consumption so small, that they can be run continuously in any reasonably well-ventilated room. It was then uprated to 60 watts ac, as described in the section 'Operating Experience', so as to enable it to power a large rotating marine light. In addition to this work, a new propane burner had to be designed, as the original burner would not provide enough power at a high enough temperature. By Spring 1977 a prototype burner was built and tested on the 60-watt TMG, giving the performance shown in Table 6.2.

This TMG was subsequently re-engineered for the Commissioners of Irish Lights. It powered their major lighthouse on the island of Eeragh, in Galway Bay, from August 1978 (Figure 6.7). The light was put under photoelectric control in November, and the crew left just

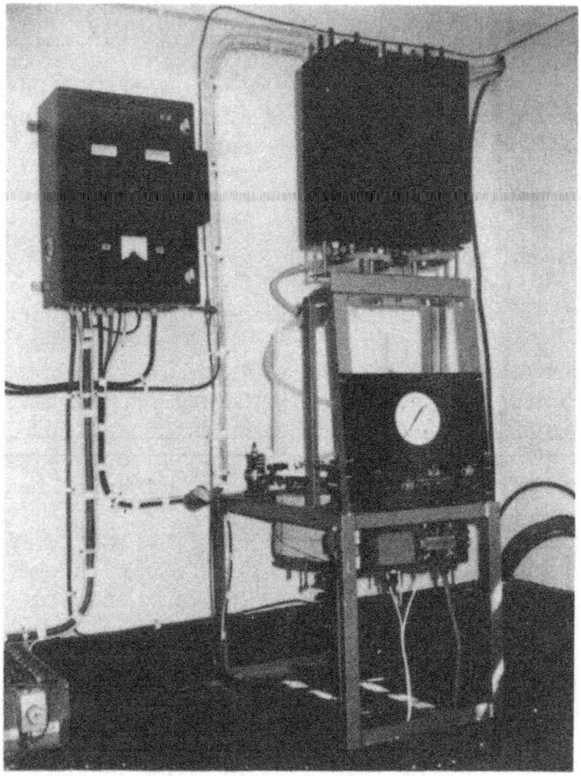

Figure 6.7 60-watt propane-heated TMG installed at the major lighthouse on the island of Eeragh, in Galway Bay, (after Cooke-Yarborough *et al* (1980).

before Christmas. Thereafter the system ran unattended, except for monthly visits. The TMG successfully powered the light through mid-winter, without requiring any help from the back-up diesel generator (Cooke-Yarborough *et al* (1980a)). The burner jet had to be replaced twice, because of slight obstruction. After the second occasion in July 1979, the propane pressure was set too high, so the temperature cut-out operated and the diesel generator took over for a few days until the next monthly visit.

After more than a year, the TMG stopped abruptly in early September 1979 because all the helium had departed. Some of the lower melting-point constituents of the braze attaching the burner to the hot end of the cylinder had segregated and attacked the stainless steel, penetrating the cylinder end. This braze had originally been intended for operation at 450°C, so failure was not unexpected. The TMG was temporarily taken out of service and the burner and cylinder assembly were remade with a higher-temperature braze. It went back into service in July 1980, and powered the Lighthouse until February 1982.

The copper heat transfer surfaces of the burner have to be coated with braze to protect them from oxidation. The coating of the original burner was found to be quite intact after a year's operation, but it appears that the remade burner was much less satisfactorily protected, for it began to deteriorate quite rapidly. The reason for shutting down in 1982 was the almost complete disintegration of the burner, though the TMG was even then still delivering about 45 watts DC. Other problems which arise were burning of the ignitor electrode, which was located directly in the flame, and intermittent blockage of the propane jet. All these problems are dealt with in later designs.

CONCLUSIONS

The field trial generators and the radioisotope generator had together totalled 126,000 machine hours of operation (equivalent to 14 years) by April 1983. None of these machines suffered any failure of the critical flexing components (the diaphragm and the displacer spring).

Most of the problems which did arise during the field trials were unrelated to the use of moving parts in the TMG. They were quite rare, and became rarer as the field trials proceeded, as is shown by comparison of the first Data Buoy mission with the second. The sealing problems need not have arisen had the alternator space been designed originally to contain helium, and had the 60 watt machine been designed originally for operation at 600°C. Burner-jet obstruction would occur equally in a propane-heated thermo-electric system. A simple mechanical device for automatically clearing the jet in situ, has since been developed by HoMach Systems Ltd. to overcome this problem.

The radioisotope heat source functions without problems. This is probably the most reliable heat source conceivable. In situations where the need for reliability, or the need to operate without air, outweighs cost considerations, the use of radioisotope heat should certainly be considered.

The TMG has been shown to be capable of providing electrical power of a few tens of watts with a fuel consumption much lower than that of a thermo-electric generator, and with good reliability. In field trials extending over five years, unattended operation over many months has been achieved. Minor improvements have since been devised as a result of this experience. These should extend the period between maintenance visits to a year or more. There is no indication that there is any limit to the operating life of the TMG.

NOMENCLATURE

A_d	cross-section area of displacer
A_m	cross-section area of permanent magnet
A_o	effective area of diaphragm
B	flux density in permanent magnet
c	ratio of outer to inner radius of diaphragm
i	instantaneous coil current in alternator
M	total engine body mass
M_o	total mass associated with diaphragm hub
N	total number of turns on alternator windings
P	mean pressure of working gas
S_d	stiffness of displacer springs
S_f	stiffness of feedback spring between hub of diaphragm and displacer
S_t	total of all the stiffnesses associated with the diaphragm hub
T_c	wall temperature of compression space
T_e	wall temperature of expansion space
V_e	volume swept by displacer
V_h	gas volume in space with wall temperature T_e (with displacer at midstroke)
V_m	total volume of working gas (when diaphragm is at mid-stroke)
V_o	volume swept by diaphragm
ω	angular operating frequency (radians per sec)
X	width of each alternator air gap (at mid-stroke)

REFERENCES

Cooke-Yarborough, E.H. (1967). The Thermo-mechanical Generator: A Proposal for a Heat-Powered Non-Rotating Electrical Alternator. Harwell Memorandum No. 1881.

Cooke-Yarborough, E.H. (1968). Diaphragm Design for a Thermo-mechanical cal Generator. Harwell Memorandum No. 2021.

Cooke-Yarborough, E.H., Franklin, E. and Geisow, J. (1968). The Estimated Power Output and Efficiency of a Thermo-mechanical Generator. Harwell Report No. 5750.

Cooke-Yarborough, E.H., Franklin, E., Geisow, J. and Howlett, J. (1973). An Assessment of the Performance of the First and Second Experimental Versions of the Harwell Thermo-mechanical Generator. Harwell Report No. 7265.

Cooke-Yarborough, E.H. (1974). Fatigue Characteristics of the Flexing Members of the Harwell Thermo-mechanical Generator. Harwell Report No. 7693.

Cooke-Yarborough, E.H., Franklin, E., Geisow, J., Howlett, J. and West, C.D. (1974). The thermo-mechanical Generator for An Efficient Means of Converting Heat to Electricity at Low Power Levels. Proc. I.E.E., Vol. 7, pp. 749-751.

Cooke-Yarborough, E.H. (1974). Simplified Expressions for the Power Output of a Lossless Stirling Engine. Harwell Memorandum No. 2437.

Cooke-Yarborough, E.H. and Yeats, F.W. (1975). Efficiency Thermo-mechanical Generation of Electricity from the Heat of Radio-isotopes. Proc. 10th I.E.C.E.C., p. 1003, Delaware, Aug. 17-22.

Cooke-Yarborough, E.H. (1976). Improvements in or Relating to Resilient Coupling Devices. British Patent No. 1,539,034.

Cooke-Yarborough, E.H. (1977). Calculated Increase in Power Output Obtainable by Increasing the Working Pressure of a Thermo-mechanical Generator Using the Present Design of Corrugated Diaphragm. Harwell Memorandum No. 2885.

Cooke-Yarborough, E.H. (1977b). A Data Buoy Powered by a Thermo-mechanical Generator. Proc. 12th I.E.C.E.C., Washington, D.C., pp. 1370-1377.

Cooke-Yarborough, E.H. and Spottiswoode, N.L. (1980). A 60-Watt Thermo-mechanical Generator as the Main Power Source for a Major Lighthouse. Proc. Xth I.A.L.A. Conf., Tokyo, Section 5.4.2.

Cooke-Yarborough, E.H. (1980). Small Stirling-Cycle Power Sources in Marine Applications. Proc. Oceans 80 Conf., Seattle, pp. 457-462.

Cooke-Yarborough, E.H. (1982a). Diaphragm Stirling Engines: Achievements and Future Potential. Proc. I. Mech. E. Conference "Stirling Engines - Progress Towards Reality" at University of Reading, pp. 43-48.

Cooke-Yarborough, E.H. (1982b). A Way to Relax the Dimensional Tolerance Requirements for Clearance Regenerators. Proc. 17th I.E.C.E.C., Los Angeles, pp. 1835-1837.

Franklin, E. (1968). Pressure, Mass Distribution and Velocity of the Gas in a Thermo-mechanical Generator Stirling Cycle Engine. Harwell Memorandum No. 2106.

Franklin, E. (1970a). An Assessment of the Suitability of the Inductor Generator for Use with the Thermo-mechanical Generator Stirling Cycle Engine. Harwell Memorandum No. 2270.

Franklin, E. (1970b). Heat Regenerator Design for the Thermo-mechanical Generator Stirling Cycle Engine. Harwell Memorandum No. 2320.

Franklin, E. (1971). Comparison of the Thermal Conduction Losses in the External and Internal Displacer Thermo-mechanical Heat Engines. Harwell Memorandum No. 2476.

Howlett, R. (1970). The Computed Performance of a Thermo-mechanical Generator Undergoing Sinusoidal Motion. Harwell Memorandum No. 2345.

Ryden, D.J. (1976). Unpublished supplement to: Energy Losses Due to Irreversible Gas Processes in Fluidyne Pumps. Harwell Report No. 8291.

West, C.D., Franklin, E. and Geisow, J. (1968). Gas Flow Losses in a Thermo-mechanical Generator. Harwell Memorandum No. 2089.

West, C.D. (1969). Diaphragm Calculations and Design for the Thermo-mechanical Generator. Harwell Memorandum No. 2177.

West, C.D. (1970). Power Loss Due to Kinetic Energy of the Gas in a Thermo-mechanical Generator Cavity. Harwell Memorandum No. 2330.

West, C.D. (1971a). Corrugated Diaphragms for the Thermo-mechanical Generator Heat Engine. Harwell Memorandum No. 2431.

West, C.D. (1971b). Outline Design of a Thermo-mechanical Generator Using an Enclosed Displacer. Harwell Report No. 6755.

CHAPTER 7 - SMALL FREE-PISTON STIRLING ENGINE: POWER SYSTEMS FOR THE ARTIFICIAL HEART

This chapter was contributed by Dr. William Martini of 2303 Harris, Richland, Washington, 99352. Dr. William Martini, now in private practice as a consulting engineer specializing in Stirling power systems, led, for many years, one of the two successful teams concerned with development of the artificial heart.

Martini publishes the Stirling Engine Newsletter, containing items of interest to the Stirling engine community, a directory of world-wide activities in Stirling engines and the most extensive bibliography of Stirling engine publications. Martini has one of the world's leading collections of Stirling engine literature and generously makes copies of any publication available on request for a nominal charge.

Martini Engineering also publishes and sells computer programmes and manuals for Stirling engines as well as several books on different aspects of the field. They are among the leading independent consultants for Stirling engine developments in North America.

The annual InterSociety Energy Conversion Engineering Conference is the focus of publication for new developments in Stirling engines. This is due in great part to the tireless efforts of William Martini over the past 15 years or so to create and foster a sympathetic interdisciplinary niche for Stirling engines amid the engineering society structure.

G. Walker

INTRODUCTION

Small Stirling engines are made in large numbers every year. They are principally used as cryocoolers in electronic applications requiring minute refrigeration at cryogenic temperature levels. Some of these units are free-piston machines and are briefly discussed in Chapter 9 and in more detail elsewhere (Walker (1983)).

Other small Stirling engines are made by model makers and hobbyists principally for recreation. Very high standards of ingenuity, resource, engineering design and craftsmanship are invested in these machines and their study should not be overlooked. Ross (1983), well-known in the small Stirling engine field has reviewed the model and miniature Stirling engine field (see also Chapter 20, Walker (1980)).

In addition to these two fields a significant and sustained programme of scientific research and engineering development of small Stirling engines has been carried out to develop power systems for the artificial heart. We shall concentrate our effort here in a review of the significant contributions to this program and assess their relevance to other applications.

THE ARTIFICIAL HEART

Heart attacks are a major cause of death and substantial efforts are in progress to devise machines that are helpful in the treatment of heart disease. Blood pumps are in the course of development and may be broadly divided into two classes:

a) heart assist machines,
b) total heart replacement.

Heart assist machines operate in conjunction with the natural heart and assume much of the pumping function of the left ventricle of the natural heart (approximately 80 percent of the total pump work required of the heart). Total heart replacement machines replace the heart entirely and assume the entire function of both the left and right ventricles.

The power systems used to drive the mechanical blood pumps may be external to the body or may be totally implanted. Presently available systems tend to be large, relatively complicated external devices permitting only limited mobility of the wearer. However, efforts are being made to devise systems both external and implanted to allow virtually the unrestrained mobility of the patient.

The programme of artificial heart development is being carried out in the United States by an agency of the United States Government, the National Institutes of Health, in particular the Cardiovascular Devices Branch of the National Heart, Lung and Blood Institute of Bethesda, Maryland. Annual conferences of the many contractors involved in the artificial heart program are held in Bethesda and the proceedings of the annual conference may be consulted for details and up-to-date information on the progress in various aspects.

There are many ethical, legal, social and financial questions less straightforward than the mechanical and technical problems of the artificial heart. We shall not attempt to address or even define these questions here.

STIRLING ENGINES IN ARTIFICIAL HEARTS

Small free-piston Stirling engines are among the systems under development as implantable power systems to drive the blood pumps, either total heart replacement or heart assist machines. An output in the range of 2 to 5 watts is necessary to operate the blood pumps and may be in the form of a mechanical shaft drive, or as a pneumatic or hydraulic output. With an implanted heart engine the thermal efficiency of conversion is important to minimize the power input and also because for the 'waste heat' to be dissipated from the power system it has to be absorbed first by the blood circulating in the body and eventually released to the atmosphere by breathing or transfer through the skin. A power output of 5 watts from an energy converter of 15 percent efficiency requires a power input of approximately 30 watts and a thermal dissipation of 25 watts; a significant additional load to the normal cooling load of the body (100 to 500 watts depending on physical activity).

Initially the development of thermal converters was predicted on the use of a radioisotope energy source (Plutonium 238) but strong public reaction to atomic energy generally has displaced the isotope energy source in favour of an electrically charged 'lithium fluoride' thermal battery of sufficient capacity to sustain engine operation for several hours before recharging is necessary.

Work on the artificial heart commenced in the mid-1960's and has been in continuous development since then. An evaluation of early concepts for the artificial heart (Hittman (1966)) was biased heavily in favour of electrically driven systems and in consideration of the potential heat engine systems concluded 'the Stirling cycle engine is not competitive in the power range of interest'. Despite this unpromising start the first research and development contracts placed by the National Heart and Lung Institute included two Rankine (steam) cycle and two Stirling engines. Both the Stirling engine concepts were free-piston machines and have enjoyed continuous support since that time and have reached the advanced stage described below. One of the steam engine concepts evolved in course of development into a free-piston Stirling engine and although funding has ceased for this unit a description of the system is included because of its unusual interest and potential application elsewhere.

THE AEROJET - GENERAL ENGINE

Figure 7.1 is a block diagram of the elements Aerojet-General artificial heart system. The engine module consists of a thermal source and the engine a 'thermocompressor' using heat from the thermal source to produce pneumatic power (compressed helium). This is supplied to the blood pump module comprised of the blood pump and a pneumatic actuator with a sophisticated logic controller regulating helium to the actuator.

Figure 7.2 shows the same system in considerably more detail with the engine, the principal focus of our interest located in the upper left quadrant of the figure.

It will be noted immediately that the engine has valves

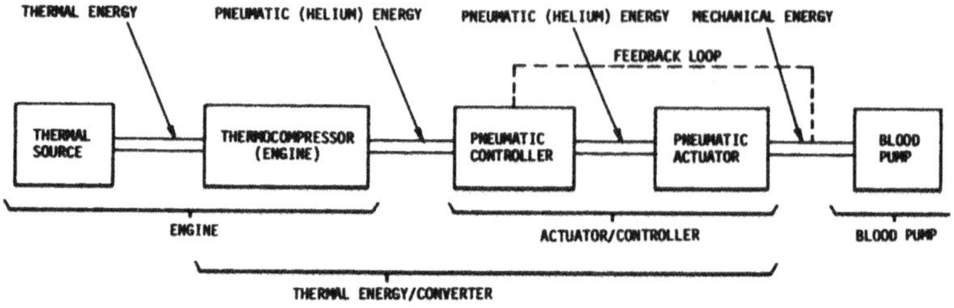

Figure 7.1 Block diagram of the all-pneumatic thermal artificial heart assist system.

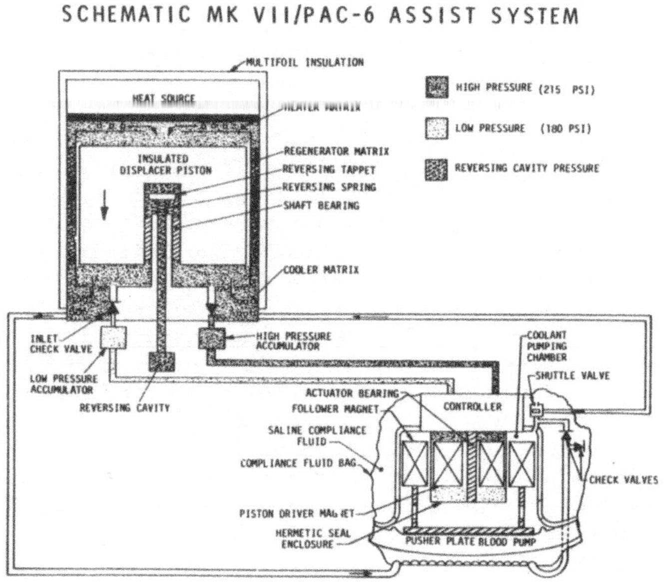

Figure 7.2 Schematic of the Aerojet-General artificial heart assist system.

regulating the inlet and outlet flow of working fluid (helium gas) to the cylinder. Under the definition included in Chapter 1 the unit is therefore an Ericsson engine rather than a Stirling engine. Nevertheless we include it here for it is an interesting machine, often miscalled a Stirling engine, and providing very severe competition indeed for the other Stirling engine artificial heart concept, the McDonnel Douglas/University of Washington engine considered below.

The Aerojet-General engine consists of an insulated displacer mounted on a centre-post and reciprocating in the cylinder.

The cylinder is heated at the top end by the heat source and cooled at the bottom end by a water cooler. Spring loaded inlet and outlet check valves are located at the lower end of the cylinder.

Operation of the engine illustrated in Figure 7.3 occurs as follows.

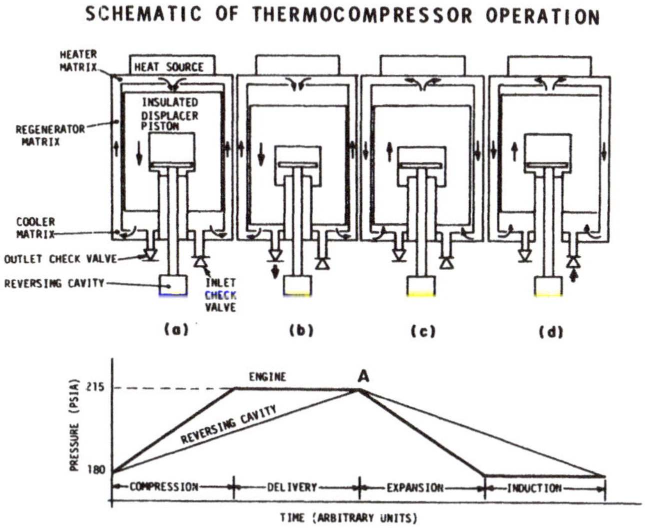

Figure 7.3 Schematic of thermocompressor operation.

As the displacer moves from top dead center (TDC) to bottom dead center (BDC) gas is displaced from the cold space through the cooler, regenerator and heater to the hot space. The gas is heated at constant volume so the pressure increases until the outlet check valve opens and gas flows from the cylinder for the remainder of the downward stroke.

Descent of the displacer causes an increase in the pressure of the helium gas spring contained in the centerpost 'reversing cavity'. This is sufficient to halt and reverse the downward motion of the displacer but is augmented by the mechanical spring and 'reversing tappet' shown in Figure 7.2.

On reversal of displacer motion the pressure falls slightly causing the outlet check valve to close. Then as the displacer ascends the working fluid passes from the hot space through the regenerator to the cold space. The mean temperature of the gas falls thereby reducing the pressure until the inlet valve opens and a fresh charge of helium is admitted to maintain a constant cylinder pressure thereafter. The centerpost gas spring pressure falls as the displacer ascends thereby increasing the volume of the reversing cavity. As the displacer approaches the top dead center position the reversing tappet is engaged causing piston reversal and the next cycle commences.

The thermocompressor is therefore a simple machine using only a displacer with two check valves and no power piston. The date of its first use is unknown but it was earlier described and illustrated in great detail by Bourne (1878) and was reinvented by Bush (1939).

Progressive development of the Aerojet-General engine led to the machine designated MK-8 illustrated in Figures 7.4, 7.5 and 7.6 with the specification given in Table 7.1 described by Andrus *et al* (1979).

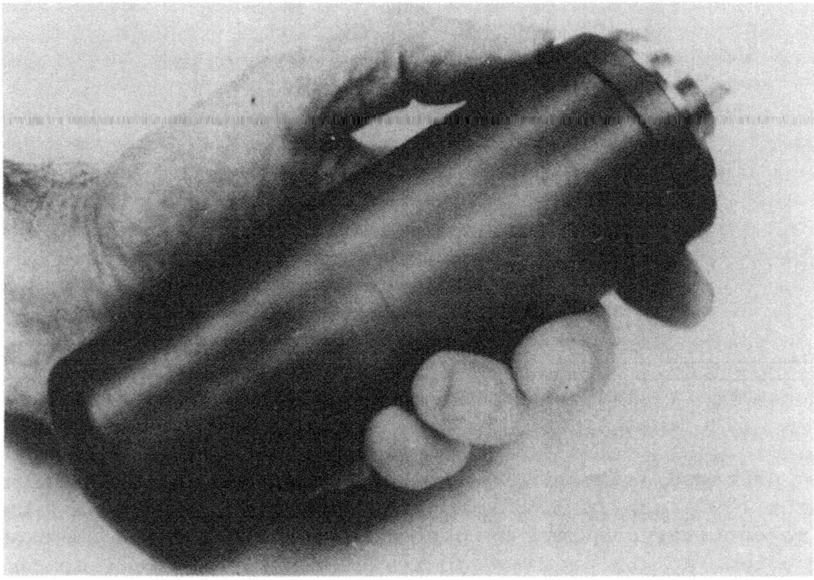

Figure 7.4 Aerojet-General MK-8 engine.

The heat source, a radioisotope capsule is contained within a cylindrical enclosure and heat is conducted from the capsule to the engine heater matrix. Multifoil superinsulation is used to thermally isolate the capsule and is extended down the length of the engine cylinder to minimize radial heat loss.

The annular regenerator contained between the cylinder wall and the displacer is made of glass 'straws' (see Walker (1980)) to minimize thermal conduction.

The bottom end cylinder head contains the centerpost on which the displacer is mounted and the inlet and outlet check valves and the water cooled heat exchanger. The gas spring and mechanical reversing spring are located within the centerpost. The high pressure and

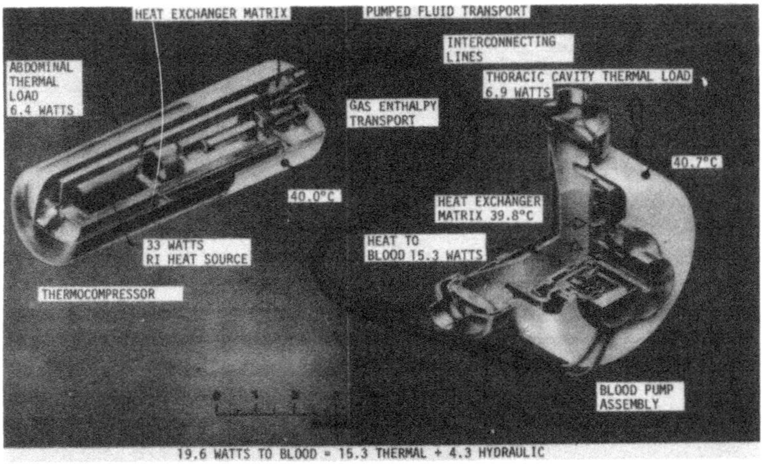

Figure 7.5 Components of MK-8 Aerojet-General engine.

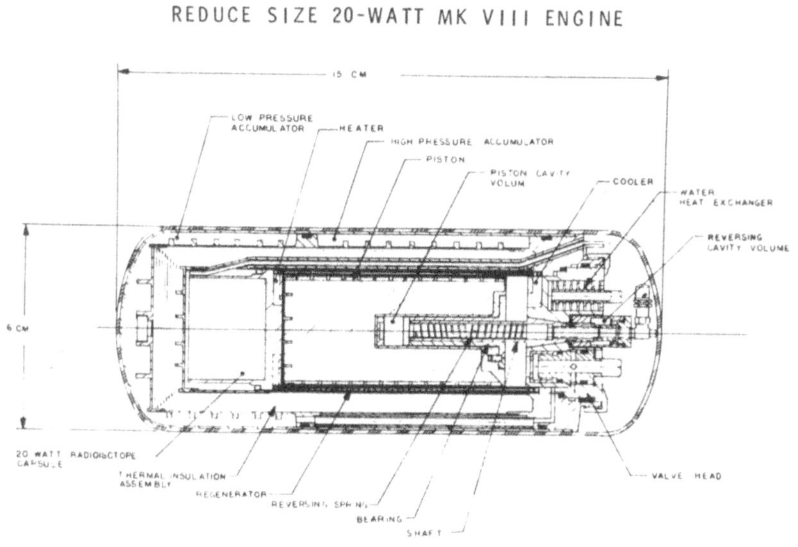

Figure 7.6 Aerojet-General MK-8A engine assembly.

low pressure gas accumulators form an annulus around the cylinder.

Further development of a smaller, lighter engine with a reduced power output was anticipated by Andrus *et al* (1979). The new machine designated MK-8 is illustrated in Figure 7.7 with the calculated energy distribution shown diagrammatically in Figure 7.8.

Table 7.1
Specification of the Aerojet General MK-8 Engine

	MK-8
Radioisotope Power Level - watts	20
Steady-State Engine Efficiency - %	17.0
Length - cm	16.5
Diameter - cm	6
Volume - cc	430
Weight - gms	940
Specific Gravity	2.18
Design Life - years	10

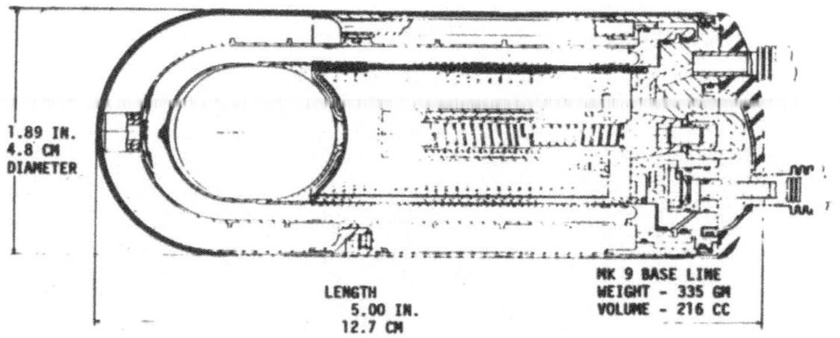

Figure 7.7 Aerojet-General MK-8 baseline engine.

More radical changes under consideration include:

a) displacer support on external bearings rather than the center-post to eliminate a difficult alignment problem,

b) use of a crankshaft and flywheel to permit the use of counter-weights with improved balancing, a slower cycle rate without stalling and silent operation by elimination of 'bang-bang' operation of the displacer as it impacts the reversing tappet at the end of each stroke.

Advantages

The principal advantage of the free-piston thermocompressor is simplicity with relatively few moving parts and with reasonable efficiency provided careful attention is paid to thermal design.

A pneumatic output can be an advantage or disadvantage depending on the application. If a pneumatic output is not specifically desired some form of expander is necessary to realise a mechanical work

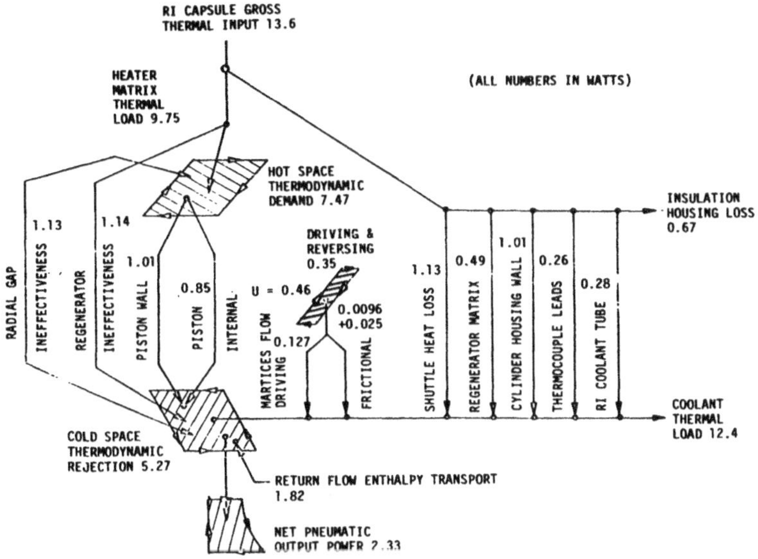

Figure 7.8 Energy flow of MK-9 Aerojet-General engine.

output and the introduction of this could result in a significant decrease in efficiency.

Disadvantages

The principal potential weaknesses of a thermocompressor in the Aerojet-General form are the valves, the dry lubricated sliding bearing of the centerpost/displacer and the mechanical springs. Detritus resulting from bearing wear could lodge on the faces of the valve disks to prevent closure. Springs, used on the valves and for the reversing system, may have a finite life depending on design and may fail by fatigue eventually. The requirement for high heat transfer but low load volume and low flow resistance is more severe than for the normal type of Stirling engine.

Use of a low clearance gas bearing to eliminate detritus is possible but extremely fine fits with matched pairs are necessary to maintain the working space/reversing cavity seal with minimal friction. Very high standards of concentricity are necessary for proper operation.

Other Applications

There appears to be no intrinsic reason inhibiting application of the thermocompressor in other applications. This type of thermocompressor should not be confused with the one used in cryogenic coolers. One possibility is used as the thermocompressor in a gas-fired Rankine cycle refrigerator or heat pump system with Freon or other working fluid. No studies of this application are known in the available literature. This could be a hermetically sealed unit but multiple compressors would be needed to generate the desired pressure ratio.

THE JCGS ENGINE

Development work on this engine started simultaneously with the thermocompressor discussed above. The initial development was done in the laboratories of the McDonnel Douglas Company and was later transferred to the Joint Center for Graduate Study (JCGS) of the University of Washington, Richland, Washington.

The initial machine was a single-cylinder free displacer unit with the pressure variations acting directly upon the diaphragm of the blood pump. The displacer drive was a novel 'plug-in-orifice' type. This concept was subsequently abandoned. There seemed little possibility that the size of engine could be reduced sufficiently to be implantable.

The second unit was a thermocompressor thought to be a reinvention along with the Aerojet engine, of the old Bourne/Bush engines. The original displacers were porous incorporating a regenerator but later became hollow structures with an annular regenerator.

Subsequently the engine was converted from an Ericsson machine with valves, the thermocompressor producing a pneumatic output, to a Stirling engine without valves in the gas. Cyclic variation of the working fluid pressure was deployed on a corrugated diaphragm to generate a hydraulic output.

Figure 7.9 is a block diagram of the elements of the JCGS engine and Figure 7.10 is a line diagram showing the system in some detail. As before the system comprises two principal assemblies, the engine module and a blood pump module.

The engine module incorporates:

a) a heat source, either an electrically heated thermal battery or a radioisotope capsule providing heat to the hot space of the Stirling engine,

b) the Stirling engine,

c) the hydraulic converter/accumulator.

A cross-section of the System 6 engine module described by Johnston *et al* (1980) is shown in Figure 7.11.

Basically the unit is a free-piston Stirling engine similar

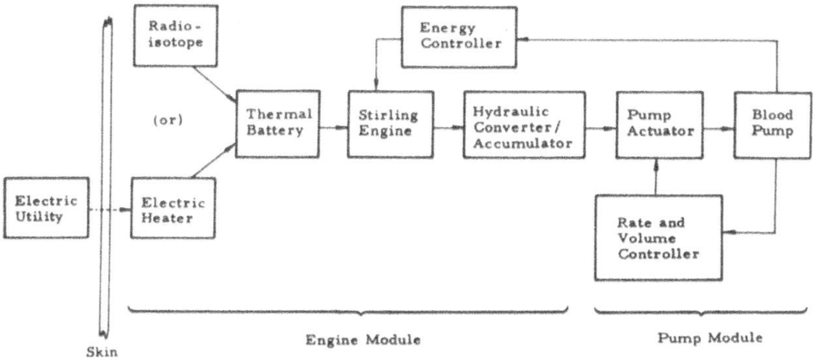

Figure 7.9 Block diagram of JCGS artificial heart system.

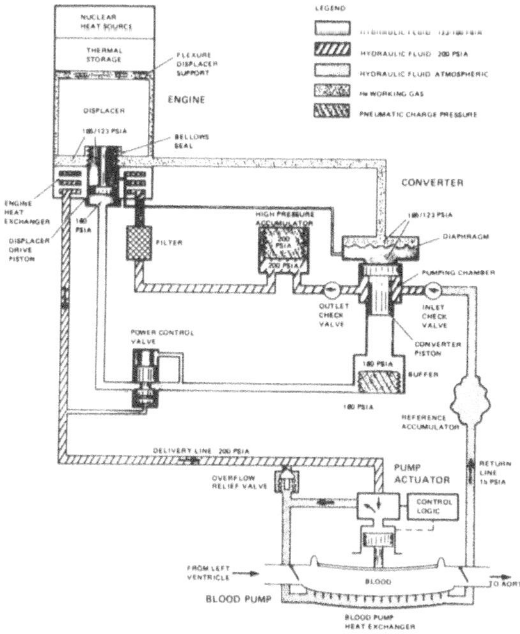

Figure 7.10 Schematic diagram of JCGS artificial heart system.

to that pioneered by William Beale at Sunpower. The displacer oscillates in the cylinder causing cyclic variation of the working fluid pressure. This acts on the 'piston', a corrugated metal diaphragm located at the bottom of the adjacent cylinder. Flexing of the diaphragm in response to the pressure variation causes an increase in the pressure of hydraulic fluid above the diaphragm and a Bramah hydraulic

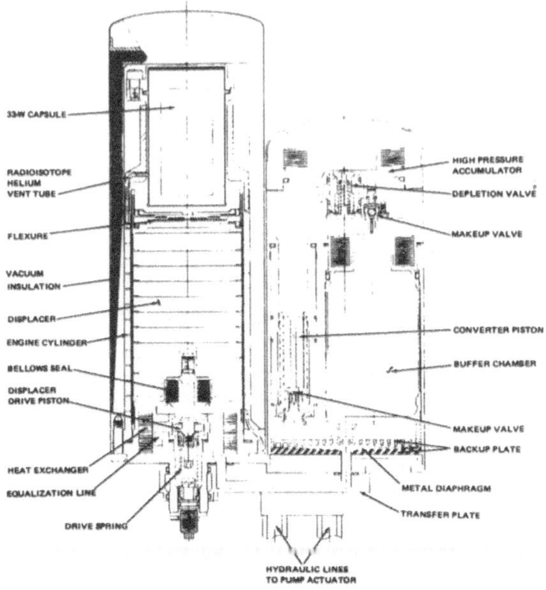

Figure 7.11 Engine module of JCGS system 6 engine.

magnifier operates to generate a high pressure hydraulic output. The system includes a hydraulic accumulator incorporating a gas filled metal bellows to provide a reserve of hydraulic power.

The engine is virtually self-starting. From a cold start the expansion space is heated by the source and the displacer becomes unstable so that any minor and random perturbation will cause it to start oscillating. As it moves down the cylinder gas will be displaced from the cold to hot parts causing an increase in the pressure and accelerating the displacer. The motion of the displacer is resisted by the spring rate of the metal bellows seal below the displacer and coupling it to the displacer drive cylinder. However the spring rate is very small and insufficient to prevent the displacer moving to the lower-most point of its stroke. Downward motion of the displacer increases pressure of the working fluid and this causes the converter diaphragm to flex and cause the converter piston to move and pump a small amount of liquid to the pumping chamber of the high pressure accumulator. Inertia of the converter piston carries it past the equilibrium point and so depressing the pressure of the working fluid and causing the displacer to ascend along the cylinder to the top of its stroke and commence a new cycle.

A unique feature of the engine is the speed control valve. If the valve is closed the engine stops but will restart if the valve is opened. The engine starts slowly and accelerates rapidly to a steady operating frequency depending on how much the speed control valve is opened.

Other novel features of the engine include the cruciform flexure at the upper, hot, end of the displacer. This guides the displacer top end for the flexure is very stiff in the radial direction but offers virtually no resistance to motion in the axial direction. This is a feature that could be deployed with advantage in many larger Stirling and Vuilleumier cycle engines.

Another feature of unusual interest in the JCGS engine is the use of metal bellows and diaphragms and their development to long-lived status. Test lifetimes of 4 and 5 years have been routinely established. Provided the maximum stress is well below the fatigue limit and meticulous attention is given to fabrication, welding and quality control the possibility has been demonstrated time and again for bellows and diaphragms to have reliable operation over many years. In the present engine ingenious design permits the bellows to act simply as barriers separating one fluid (helium) from another (hydraulic) with no significant pressure difference between them. In this situation bellows failures are rare and if they do occur they arise as a consequence of contamination or improper fabrication not discovered by inspection. However, it must be said that the greatest system reliability was achieved with the System 4, a relatively simple machine incorporating a crank-driven displacer and a metal bellows under some pressure stress. Built in the early 1970's the unit achieved over 4 years cummulative operation. It is of interest to note the Aerojet-General team project the use of a crank-driven unit to reduce the noise and vibration of their free-piston unit.

Table 7.2 contains the performance of System 6 and System 7 JCGS Stirling engines. The data given for the System 6 engine is actual measured data from a prototype unit. The data for System 7 is the projected design performance intended to develop a reduced power output specified by the National Institute of Health.

Figure 7.11, a cross-section of the System 6 JCGS engine is also representative of the System 7 engine, virtually a scaled version of System 6.

White *et al* (1982a) gave a progress report or general review of the recent JCGS engines including a summary of the computer simulation programmes available for design support and optimization.

Potential Applications

Robust efforts are being made by the JCGS team to extend and apply the specialized technology developed for the artificial heart engine. White *et al* (1982b) summarized a design study for a 15 kW free-piston Stirling engine with hydraulic output. The study was carried out for the National Aeronautics and Space Administration (NASA) Lewis Research Centre for a solar power/hydraulic output free-piston Stirling engine. A Sunpower RE 1000 free-piston Stirling engine

Table 7.2
Power Source Design and Performance Parameters

Prototype Designation	Documented System 6 Implantable	Pending System 7 Implantable
Thermal Engine Module Characteristics		
Hydraulic power to actuator (w)	5.76	5.0
Working fluid - engine	Helium	Helium
- hydraulic converter	NF No. 1	NF No. 1
Peak operating frequency (Hz)	25	35
Peak cycle temperature (°C)	575	500
Cycle pressures (psia) - engine	200/160	230/155
- converter	215/14.7	250/14.7
Peak engine module efficiency (%)	17.5	25**
Fueled weight (kg)	1.80	0.6
Volume ( )*	0.77/1.05	0.25/0.33
Key dimensions (cm)		
Maximum length*	16.3/17.3	11.4/12.1
Maximum width*	10.2/12.1	7.1/8.5
Maximum thickness*	5.1/7.2	3.6/5.0

*With no low temperature foam insulation/with insulation.

**20% efficiency so far observed.

was specified as the power unit so that the JCGS team were unable to fully deploy the full range of their novel approaches and undoubted design expertise. Nevertheless the paper makes very stimulating reading and promises well for the future. A parallel contract for the hydraulic output free-piston Stirling engine was reported by Harvel *et al* (1982).

COMPARISON OF THE JCGS AND AEROJET GENERAL ENGINES

Comparison of Figure 7.11 and the equivalent diagram of the Aerojet General engine, Figure 7.6, reveals that the thermocompressor is considerably smaller and less complicated than the JCGS engine. However the JCGS engine has achieved significantly better performance, a 20 percent conversion efficiency compared with a maximum of 17 percent for the Aerojet-General engine. Moreover, there are those who hold that the Stirling engines are intrinsically more efficient that thermocompressor engines and that substantial development potential remains in the JCGS engine whereas the Aerojet-General machine has already approached the limit of development. Another significant difference in the two machines is the output, pneumatic in one case, hydraulic in the other. The Aerojet-General machine has an apparent advantage that the working fluid, helium, is used to actuate the blood pump with no need for a hydraulic converter. However, this is largely negated by the need for a liquid cooling system on the engine coupled by hydraulic line to the blood pump for the thermal dump. On the JCGS engine the hydraulic connection serves both for the power transmission and coolant coupling function. The need for both hydraulic and pneumatic coupling of the engine and pump module necessary with the Aerojet unit is eliminated on the JCGS machine.

THERMO-ELECTRON FREE-PISTON STIRLING STEAM ENGINE

The Thermo-Electron Corporation of Waltham, Massachusetts was among the pioneer companies selected to evaluate the application of heat engines to the artificial heart. The company initially deployed their special expertise of conventional reciprocating steam engine technology adapted to miniature engines. In the early 1970's, however, the company developed the *'tidal regenerator engine'* virtually indistinguishable from a Stirling engine with a condensing working fluid and with the significant advantage of operating at heart beat frequency. Watelet *et al* (1976) summarized the development of the engine. Figure 7.12 is a diagrammatic cross-section of the tidal regenerator engine. Heat is supplied both to the superheater and, principally, to the boiler. The boiler heat input first passes through thermoelectric modules to develop electric power for the electronic controls and the DC torque motor driving the piston. The electronic controls sense that the blood pump requires pressurized fluid from the engine and so cause the DC torque motor to elevate the 'piston' driving water from the 'compression space' to the displacer cylinder. As the liquid level in the condenser is raised into the regenerator space water boils and the pressure is increased. The vapor moves through the 'annular gap regenerator' into the superheater and further heating of the vapour occurs.

This addition of a small quantity of water generates a substantial increase in pressure as a consequence of the phase change. This causes the displacer cylinder to descend pumping liquid from the bellows below the displacer to the blood pump. The liquid leaving the bellows passes next through the condenser picking up waste heat for dissipation from the blood pump.

When the electronic control system senses that a reduction of pressure is required it reverses the torque motor, winding down the piston and causing a return flow of liquid to the motor cylinder and releasing the pressure. The liquid level in the displacer then falls below the condenser inlet and condensation of vapour occurs resulting in a further decrease in the pressure. As a consequence of the pressure decrease the displacer ascends the cylinder, liquid returns from the blood pump to the bellows and the engine is restored to its original condition ready to repeat the cycle.

Intuitively one expects the tidal regenerator engine to be sensitive to its orientation but experience has shown it will operate in any position. It is extremely sensitive to gases initially dissolved in the liquid working fluid and released during operation of the engine. The water must be carefully degassed before addition to the engine.

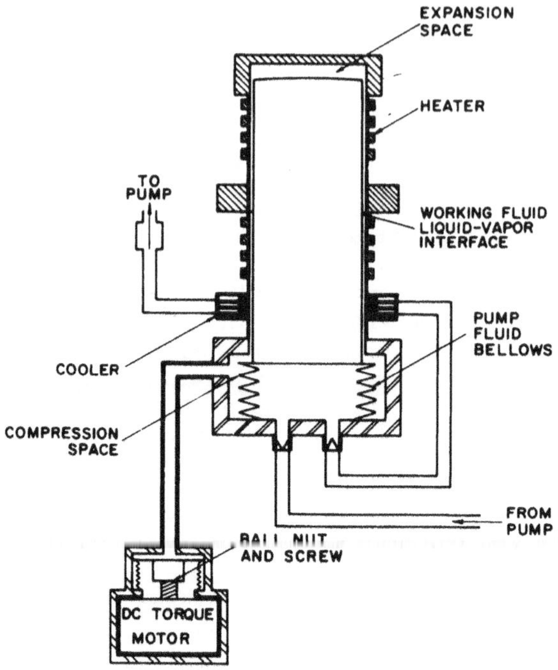

Figure 7.12a Schematic diagram for Thermo-Electron tidal regenerator engine.

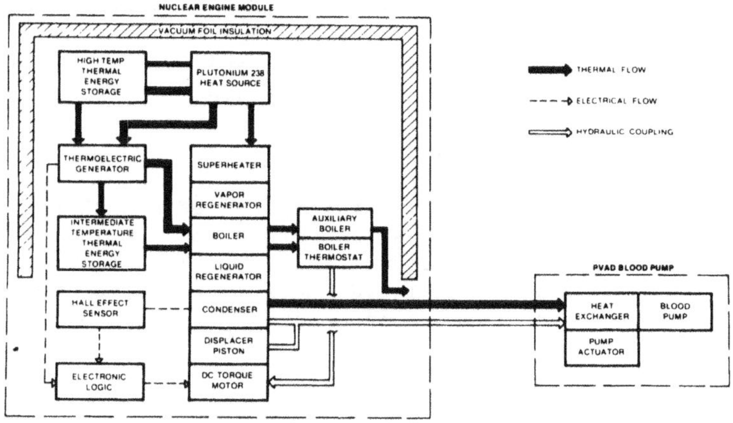

Figure 7.12b Schematic diagram for Thermo-Electron tidal regenerator engine.

Figures 7.13 and 7.14 show a cross-section of the tidal regenerator and its performance. The engine module was 20 cm (7.9 in.) long by 6.1 cm (2.4 in.) diameter, weighed 1.6 kg and occupied 0.7 liter. In conjunction with a blood pump, not shown, the unit pumped 10 liters of blood a minute against a peak systolic pressure of 160 mm of mercury.

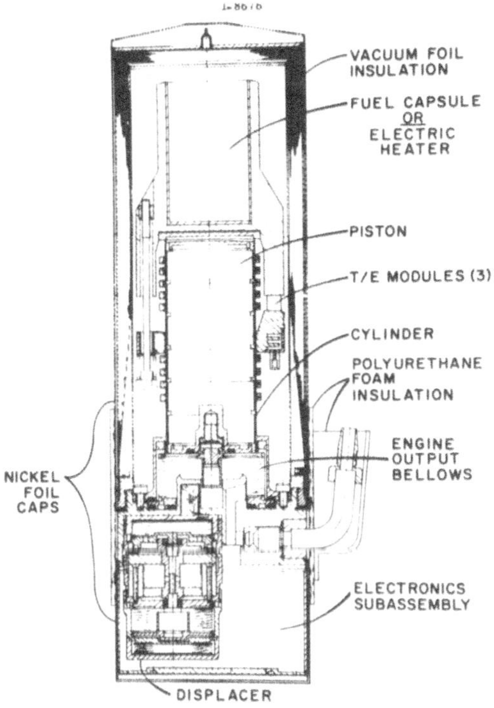

Figure 7.13 Model 4 Thermo-Electron tidal regenerator engine.

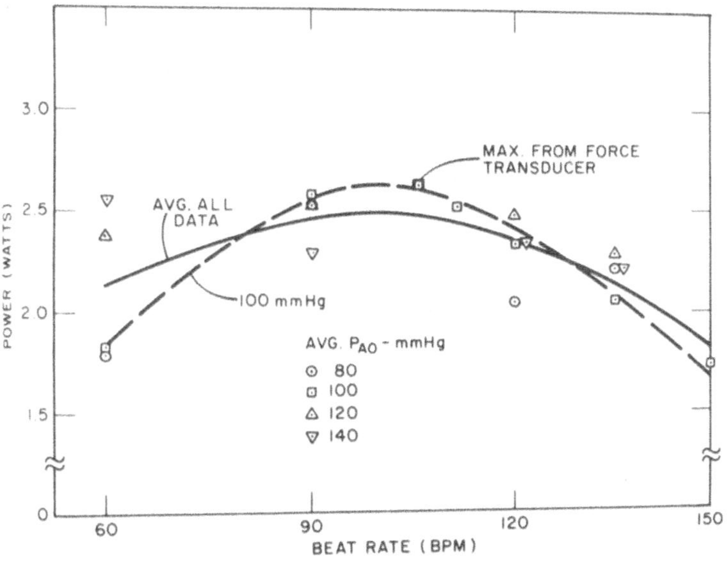

Figure 7.14 Performance characteristics of the model 4 Thermo-Electron tidal regenerator engine.

The thermal efficiency of 6 to 8 percent was significantly lower than the 15 percent of the JCGS engine and the 11.6 percent of the Aerojet General engine evaluated on the same basis and at the same time.

In an effort to improve efficiency the Thermo-Electron team developed a binary system using a diphenyl working fluid in a high temperature stage with water in the low temperature stage (see Watelet *et al* (1976)). Financial support for the work was discontinued in the late 1970's by the National Institute of Health and no further work utilizing the concept has been reported in the literature.

The tidal regenerator engine is an interesting concept but somewhat difficult to classify. In conventional Stirling engine technology the reciprocating element in the heated cylinder is called the displacer and is characterized as a zero work element. Work to and from the engine is transferred through the piston operating in the cold cylinder. The tidal regenerator engine operates such that the major work output of the engine comes from the reciprocating element in the hot cylinder, the displacer and a minor work input is supplied to the 'piston' of the cold space. Some people argue convincingly that the tidal regenerator engine should be simply classified as a 'two-piston' Stirling engine.

It is a relatively simple uncomplicated engine capable of converting a minor work input along with a major thermal input to a hydraulic output with a somewhat disappointing efficiency but perhaps capable of substantial improvement with better regenerative action. However, there is a basic problem with full regeneration. The boiling takes place at a higher temperature than the condensation.

CONCLUSION

Research and development effort on the artificial heart has been in progress for nearly 20 years. A small fraction of that effort has been devoted to exploring the use of miniature heat engines. From this work two thermal regenerative engines with a gaseous working fluid have emerged as the clear favourites for adoption with a radioisotope power source.

The use of radioisotopes is not acceptable in the popular view and the alternative of an electrically heated 'thermal battery' is being considered.

The progress achieved with the artificial heart is such that the socio-economic and legal questions appear more formidable than the techno-medical problems remaining.

On their way the Aerojet-General and the JCGS teams of highly competent engineering scientists have confronted and overcome many problems found in Stirling engines at higher power levels or used as refrigerators and heat pumps. Extensive reports of their work have been presented in the engineering and bio-medical literature. The quality of their work warrants the closest scrutiny of all those working in the Stirling engine field.

REFERENCES

Andrus, S. and Moise, J. (1979). Development and Evaluation of a Pneumatic Left Ventricle Assist Thermal Power System. Annual Report Nat. Inst. of Health, Report No. 1-HV-3-2930-5, Aug. (see also, Moise, J.C. and Faeser, R.J. (1977). Thermocompressor Powered Artificial Heart Assist System. Proc. 12th Inter. Soc. En. Conv. Eng. Conf., Paper No. 779017, pp. 112-118, Washington, D.C., Aug. 28/Sept. 2.).

Bourne, J. (1878). Examples of Steam, Air and Gas Engines of the Most Recent and Approved Type. Longmans Green and Co., London.

Bush, V. (1939). Apparatus for Compressing Gases. U.S. Patent 2157229, May.

Johnston, R.P. *et al.* (1980). Implanted Energy Conversion System. Annual Report, National Institute of Health, No. 1-HV-92908-1, July.

Ross, A. (1983). Stirling Engine Work. Live Steam, Vol. 17, No. 1, pp. 10-17, Jan.

Walker, G. (1980). Stirling Engines. Oxford University Press.

Walker, G. (1983). Cryocoolers. 2 Vols., Intl. Monographs in Cryogenics, Plenum Press, New York.

Watelet, R.P., Ruggles, A.E. and Tort, R. (1976). Status of the Tidal Regenerator Engine for Nuclear Circulatory Support System. Proc. 11th Inter. Soc. En. Conv. Eng. Conf., Paper No. 769022, pp. 136-142, Lake Tahoe, Nev., Sept. 12-17.

CHAPTER 8 - LARGE FREE-PISTON STIRLING ENGINES

INTRODUCTION

The word 'large' applied to Stirling engines means different things to different people. An engine having a 3 kW output would be large to those working on artificial hearts or cryocoolers for electronic applications. It would be small, negligibly so, to those interested in locomotive propulsion. Reference here to large Stirling engines can therefore only mean, in the most general way, engines of a size and capacity greater than is customary (recognizing of course that, apart from cryocoolers, there are no customary Stirling engines; all the machines known are 'specials' made a few at a time).

The dramatic rise in oil prices in the 1970's resulting in the emergence of natural gas as a primary fuel and the resurgence of interest in coal provide many opportunities for the application of large Stirling engines for power, cryocooling and heat pumps.

FEASIBILITY OF LARGE STIRLING ENGINES

There appears to be no intrinsic maximum size or power limit for Stirling engines, either free-piston or kinematic machines.

Gedeon (1981) formulated the scaling rules for Stirling engines. The basic geometry of an existing small or intermediate capacity engine of successful design may be 'scaled up' according to the rules proposed by Gedeon to obtain a first-order estimate of the geometric parameters for a large engine. The rules are sufficiently general that various design constraints may be introduced, speed, pressure, stroke-bore ratio, etc. and their effects evaluated. The scaling rules apply to free-piston, Ringbom and kinematic drive engines.

APPLICATIONS OF LARGE STIRLING ENGINES

Power Systems

Power applications of coal-fired Stirling engines are foreseen in the power range 0.1 to 5 MW for stationary power generation, marine propulsion, locomotives and the large off-highway vehicles used for mining, construction and forestry (Walker *et al* (1983)). All these would likely combine a fluidized bed coal combustor with a sodium heat pipe to transfer the heat of combustion to the engine.

The Stirling engines used for these applications could be free-piston, hybrid-Ringbom or kinematic drive units. Various concepts involving the conversion of large diesel engines and air compressors to operate as large coal-fired Stirling engines are being explored (Walker (1983)) as well as novel 'Scotch yoke' engines. All involve a 'free' Ringbom displacer or the Martini independently-driven displacer arrangement.

For large engines it is anticipated that air will be used as the working fluid rather than the hydrogen or helium necessary for high speed, high power density engines. The use of air is preferred to overcome the acute sealing problems that have inhibited commercial application of helium/hydrogen engines. Air is easier to contain and if leakage does occur it can be readily replenished from the atmosphere. In large, slow speed, engines having a relatively low power density, the light gases have no advantage over air as the working fluid.

Leakage of oil lubricant from the crankcase to the working space could combine with the compressed air working fluid in potentially explosive proportions especially in the heated parts of the engine. For this reason water is proposed as the lubricant for air engines. The lubrication requirements of Stirling engines are less arduous than for diesel engines. Furthermore, in most bearings the lubricant supplied acts primarily to cool rather than lubricate.

A proof-of-principle engine development of a water lubricated, Ringbom-Stirling air engine based on the conversion of a commercial diesel engine has been carried out at the University of Calgary (Walker (1983)). A cross-section of the engine is shown in Figure 8.1. Further work to produce a fluidized bed coal combustor and the associated sodium heat pipe for this engine are in progress.

Few studies of large free-piston Stirling engines have been made. Srinivasan *et al* (1983) presented the study of a 1000 horsepower yard locomotive embodying a coal-fired four-cylinder free-piston Stirling engine driving a hydraulic pump to energize a hydrostatic locomotive drive.

Large Cryocoolers, Refrigerating Machines and Heat Pumps

There is every possibility the future will see the development of large free-piston Stirling cryocoolers, refrigerating machines and heat pumps. These are of particular interest in the duplex or Stirling-Stirling form wherein one Stirling engine receiving heat at high temperature produces the power necessary to drive an adjacent Stirling engine acting as a refrigerator or heat pump.

For this configuration free-piston engines are particularly favourable. The two Stirling engines can be integrated in a very compact system with a common piston and two displacers, one hot, one cold and a total absence of connecting rods, crankshafts and the like.

Figure 8.1 Ringbom-Stirling engine based on the conversion of a commercial diesel engine, University of Calgary, 1982.

The emergence of natural gas as primary heating fuel coupled with increased interest in energy conservation has focussed attention on gas-fired Stirling heat pumps. Various efforts to produce commercial versions are underway in Europe, North America and Japan.

The principle of the natural gas-fired Stirling-Stirling free-piston heat pump is illustrated in Figure 8.2. Natural gas is burnt continuously at atmospheric pressure to energize the free-piston Stirling engine producing power to drive the associated free-piston heat pump. The heat pump abstracts heat at ambient temperature from an air, ground or water source. It delivers heat at some higher temperature for space heating or for process work. Heat is also received from the power system cooling water and also from the exhaust of the combustor. The net result is that the heat received for space heating is greater by the amount lifted from the ambient temperature source than would be the case from the alternative gas-fired heater. Whether or not the increased heat gain is sufficient to justify the expense and complexity of the heat pump depends on many factors, including the

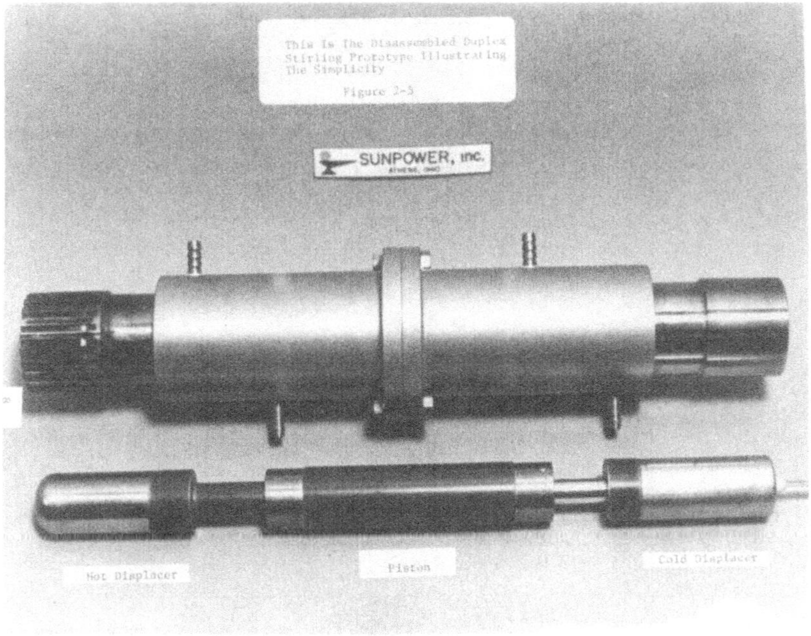

Figure 8.2 Natural gas-fired Stirling-Stirling free-piston heat pump.

price of fuel, capital costs of the equipment, depreciation and maintenance.

The Gas-Fired Natural Gas Liquefier

A need is perceived for a cryocooler having substantial refrigeration capacity (multikilowatt at 110 K) activated by thermal input. One form of gas-fired free-piston duplex Stirling cryocooler acting as a natural gas liquefier is shown in Figure 8.3. The liquefier is driven by a fraction of the gas to be liquefied and could be used to locally produce from pipeline gas the liquid natural gas fuel for transit, transport or taxi vehicle fleets, locomotives or agricultural use. Another application is for use with 'capped' gas wells to bring these hitherto undeveloped resources into use. In the search for oil many holes are drilled in potentially oil-bearing strata. Most of the time nothing is found. Sometimes natural gas is located and, very occasionally, oil, nearly always in association with natural gas. In many cases the gas does not exist in sufficient quantity to justify a pipeline connection so the well is either 'capped' or the gas is flared as oil is produced. Use of a gas-fired natural gas liquefier would facilitate development of these gas sources. A truck or skid-mounted unit would be transported to the well site and operated there to liquefy the natural gas while consuming a fraction of the gas available with no requirement for additional utility connections.

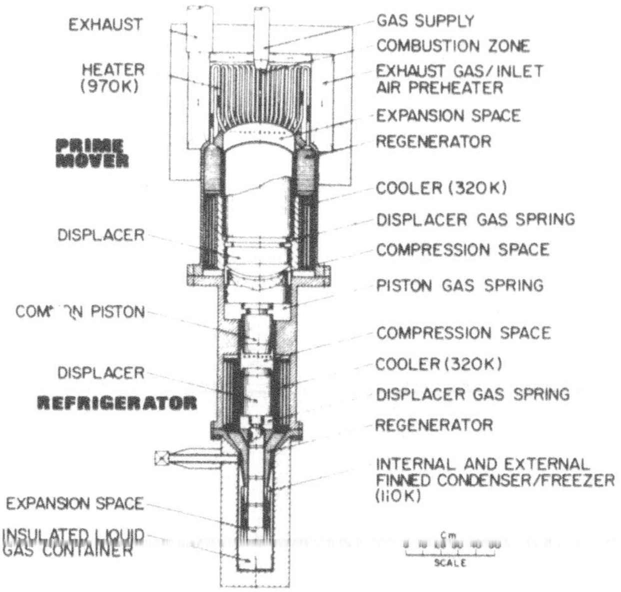

Figure 8.3 Gas-fired free-piston Stirling engine natural gas liquefier.

Unprocessed natural gas is a mixture of many components, methane, nitrogen, hydrogen sulphide, helium, water and other hydrocarbons, 'the heavy ends', propane, butane, etc. The use of switching recuperative contraflow heat exchangers in association with the duplex Stirling-Stirling liquefier could result in the liquefaction of methane and the consumption of the other combustible constituents to drive the liquefier. The unprocessed natural gas from the well would be directed through one set of flow channels in the contraflow heat exchanger. In passage through the heat exchanger it would be cooled to a condition approaching the liquefaction temperature of methane (110 K). In the cooling process many of the gas constituents precipitate on the walls of the heat exchanger so the gas passing through the exchanger becomes an increasing purified mixture of methane, nitrogen, helium and perhaps argon.

At the cryocooler the methane liquefies and is the useful product of the system. The unliquefied nitrogen, helium and argon, the 'boil-off', leave the liquefaction chamber and then pass through the return passages of the heat exchanger and increase to ambient temperature in cooling the incoming stream.

Periodically the channels are switched so that the 'boil-off' passes through the channels previously used for the incoming stream and now heavily contaminated with the condensed gas components. As the boil-off gas is heated by heat exchange with the oncoming stream the condensed fractions are re-evaporated and pass thence to the combustion chamber where they are consumed to drive the system. In this relatively simple way the heavily contaminated natural gas is 'processed', the methane component recovered and the remainder used to energize the system.

Another use for the gas-fired natural gas liquefier is to reliquefy the evaporated natural gas boil-off from methane storage tanks on board LNG tankers or land-based storage.

The results of a design study for a gas-fired natural gas liquefier for the above applications having a capacity of 20 kW at 110 K were given by Walker (1982). Berkowitz (1982) described a similar machine of lower capacity under development at Sunpower Inc., Athens, Ohio.

CONCLUSION

Although there are no large free-piston Stirling engines in production the future appears very bright indeed for such machines operating as power generators, cryocoolers, refrigerating machines and heat pumps including duplex Stirling-Stirling arrangements. Compelling advantages of the free-piston Stirling engine are that the system can be completely sealed with no piston side forces with consequent low rates of wear coupled with the use of hydrodynamic or hydrostatic gas bearings using the working fluid as the lubricant. The development of large units will undoubtedly involve large research/development expenditures which will most likely await the development and application of smaller free-piston units, less costly to develop.

REFERENCES

Berkowitz, D. (1982). The Design, Development and Performance of a Duplex Stirling Natural Gas Liquefier. Paper No. 829296, Proc. 17th Inter. Soc. Ener. Conv. Eng. Conf., Los Angeles, Aug.

Gedeon, D. (1981). Scaling Rules for Stirling Engines. Proc. 16th Inter. Soc. Ener. Conv. Eng. Conf., Paper No. 819796, pp. 1929-1935, Atlanta, Ga., Aug.

Srinivasan, V., Walker, G. and Fauvel, R. (1982). Coal-Fired Stirling Engine Hydrostatic Locomotive. Paper No. 82-WA/RT-3, Am. Soc. Mech. Eng., New York, (Wint. Ann. Mtg., Phoenix, Az., Nov.).

Walker, G. (1982). Thermally-Activated Stirling Cryocooler. Proc. 9th Int. Cryo. Eng. Conf., Kobe, Japan, May Pub., Butterworths, Guildford, Surrey.

Walker, G., Kentfield, J.A.C., Johnson, E., Fauvel, R. and Srinivasan, V. (1983). Coal-Fired Stirling Engines for Railway Locomotive and Stationary Power Applications. Proc. Inst. Mech. Eng., London, Vol. 197, No. 46.

CHAPTER 9 - FREE-PISTON STIRLING CRYOCOOLERS

INTRODUCTION

A cryocooler is a device used to produce refrigeration at very low temperatures in the cryogenic range, 0 to 120 K.

As outlined in Chapter 1, Stirling engines operate well as refrigerators. In the cryogenic range, they have become well established over the past 30 years in various military and commercial applications. Miniature cryocoolers are used extensively in infrared systems for night vision and missile guidance systems. Stirling engines are virtually unchallenged in this field and several thousand units are produced annually. It is the largest single application of Stirling engines for any purpose. Diverse other electronic and instrument applications for miniature or small Stirling cryocoolers exist or are foreseen for the near future, (see Walker (1983)). In larger sizes Stirling cryocoolers compete with Vuilleumier, Gifford-McMahon, Solvay, Postle, Joule-Thomson and Claude cycle systems for all manner of low temperature refrigeration applications including cryopumping, gas liquefaction, cryo-surgery, communications, high resolution optical systems.

FREE-PISTON STIRLING CRYOCOOLER

Stirling engine cryocoolers developed for the above applications have included free-piston engines. Virtually all the machines are of the piston/displacer type and can generally be divided into:

a) Integral Stirling machines where the displacer and piston operate in the same cylinder.

b) Split-Stirling machines where the displacer and piston operate in separate cylinders, connected by a duct which may be up to several meters long.

The cylinder containing the displacer is invariably a long thin cylinder, called the 'cold finger'. The expansion space above the displacer is the region where the useful refrigeration is generated and is called the 'cold tip'.

The infrared sensor chip is located at the tip of the cold finger and fine copper leads connect the sensor to the associated electronics used to produce a visual display on a cathode ray screen. A high vacuum insulating Dewar flask (glass or stainless steel) is located around the cold finger with appropriate windows for the sensor to 'see' through.

The space below the displacer and above the piston is of course the compression space. In a split-Stirling system the compression space is divided between the two cylinders and includes the 'dead volume' of the connecting duct. To minimize the effect of this dead volume the connecting duct is made as short as possible and of relatively fine bore tubing. However, if the tube diameter is made too small the pressure wave generated by the compressor piston is attenuated thereby reducing the refrigeration that can be produced in the expansion cylinder.

Clearly the integral-Stirling engine will likely be more effective in terms of refrigeration produced per unit of piston displacement. However the split-Stirling engine has other advantages that make it the system of choice for the miniature coolers used in infrared and other electronic applications.

By separation of the displacer cylinder from the piston a good deal of the vibration, noise, either acoustic, magnetic or electric and the hurly-burly of machine operation can be eliminated from the general environment of the sensor located on the displacer cylinder tip. This is particularly important in some applications with sensors of incredible sensitivity. The compressor and drive motor can be located in a separate compartment quite isolated from the displacer cylinder and sensor.

The other principal advantage arises because the displacer cylinder is relatively lightweight compared with total mass of the complete system. This is advantageous with gimbal-mounted equipment for night vision or guidance systems where rapid, easy movements are necessary.

INTEGRAL FREE-PISTON STIRLING CRYOCOOLER

The elements of a Beale free-piston Stirling are shown in Figure 9.1. It consists of a lightweight displacer and a heavy piston contained within the single cylinder with the expansion space above the displacer and the compression space between the displacer and the piston. The space below the piston, called the 'bounce space', contains the same gaseous working fluid used in the Stirling cycle system and acts as a pneumatic spring.

To operate as a refrigerator the system requires an input of work. This could be provided hydraulically or pneumatically but in most cases it is convenient to use a linear electro-magnetic motor to provide the work input.

Application of a force on the piston by the motor will cause the piston to descend thereby compressing gas in the bounce space. Release of the force on the piston will allow the compressed gas in the bounce space to return the piston to the original position and inertia of the piston will carry it upwards beyond the original starting position.

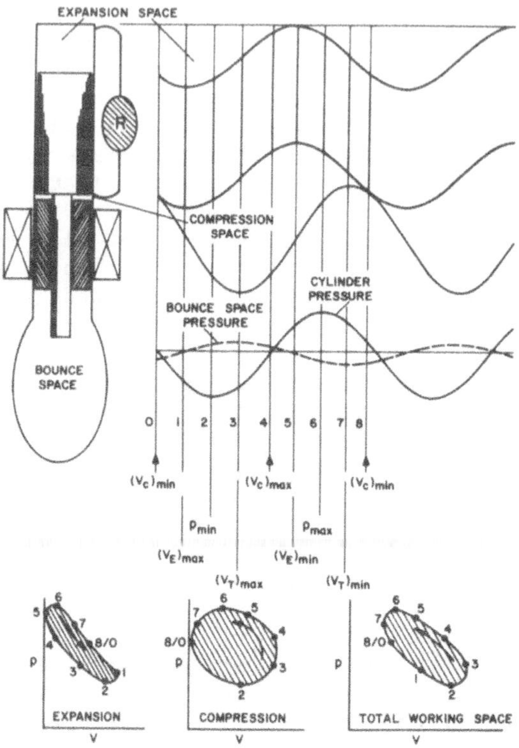

Figure 9.1 Elements of Beale free-piston Stirling cryocooler.

Application of the motor pulses at a frequency corresponding to the natural resonant frequency of the piston/bounce space will cause the piston to oscillate in the cylinder with a steady simple harmonic motion.

Oscillation of the piston in the cylinder will result in pressure of the working space varying in harmonic fashion. If the displacer were not included the space above the piston would act simply as another bounce space and the pressure variations above and below the piston would be exactly out of phase by one half (180°) and varying at the same frequency as the piston oscillation. The amplitude of the pressure variation would of course depend on the volume of the space in terms of the swept volume of the piston. The only input required to sustain this oscillation would be the minor work necessary to compensate for mechanical friction and the thermal-hysteresis losses in the two spring spaces.

To achieve refrigeration it is necessary to include the displacer and regenerative heat exchanger. In the arrangement shown the light displacer is equipped with a relatively large diameter displacer rod passing through the piston and is subject, therefore, to the pressure of fluid in the bounce space over the whole area of the displacer rod. An equivalent area on top of the displacer is subject to the pressure of working fluid in the expansion space. Thus the fluid force driving the displacer is equal to the pressure difference between the working space and the bounce space times the displacer rod area.

The *mass ratio* of the piston to displacer and the *area ratio* of the displacer rod to cylinder are critical to successful operation. When these are in proper proportion the displacer oscillates in the cylinder as shown in Figure 9.1 at the same frequency as the piston but with a phase difference of 50 to 90° ahead of the piston.

The motion of the displacer causes the working fluid to move back and forth between the expansion and compression spaces. The pressure is decreasing when the displacer is closest to the piston with the majority of working fluid in the expansion space. Therefore as the piston descends and the pressure decreases the temperature of the working fluid decreases and heat is absorbed from the surrounds of the expansion space cylinder.

Similarly the working fluid is concentrated in the compression space when the pressure is rising, with the piston near the bottom of its stroke, and ascending, with the displacer widely separated from the piston, and near the top of its stroke. As the piston rises compression of the working fluid occurs, the temperature increases and heat is rejected from the working fluid to the cooling medium, air or water surrounding the compression space or cooler.

Displacement-time characteristics for the piston and displacer and pressure-time curves for the bounce space and working space are included in Figure 9.1. Pressure volume diagrams are also included for the expansion, compression and total working spaces. It will be noted the pressure volume diagram for the expansion space is clockwise, (positive work), whereas the diagram for the compression space is anticlockwise (negative work), and larger than the diagram for the expansion space. The diagram for the total working space is also anticlockwise with an area equal to the *difference* in the area of the expansion and compression space diagrams. This is the net work input necessary to drive the refrigerator plus, of course, all the various losses, friction, hysteresis, etc. The area of the expansion space diagram is the total refrigeration generated by the system. This is particularly consumed by internal mechanical and aerodynamic friction generating heat, and by thermal conduction, convection and radiation losses so the *available* refrigeration is always substantially less than that indicated by the expansion space diagram.

In Figure 9.1 the key points of the cycle are identified by the numbers 0 through 8 in accordance with the following key:

0 - minimum compression space; the piston and displacer are in contact,

1 - maximum expansion space volume; the displacer is at the bottom of its stroke,
2 - minimum pressure in the working space,
3 - maximum system volume; the piston is at the bottom of its stroke, the volume of the bounce space is a minimum, the pressure in the bounce space is a maximum,
4 - maximum compression space volume; the displacer and piston are at maximum separation,
5 - minimum expansion space volume; the displacer is at the top of its stroke,
6 - maximum pressure in the working space,
7 - minimum system volume; the piston is at the top of its stroke, the volume of the bounce space is a maximum and the pressure in the bounce space is a minimum,
8 - minimum compression space volume; the displacer and piston are in contact as at state 0 and the cycle is complete.

Many alternative arrangements and configurations of the reciprocating elements are of course possible, a good number of which are briefly considered in Chapter 2.

One integral free-piston Stirling cryocooler in commercial production is shown in Figures 9.2 and 9.3. This unit called the M.C. 80 was developed in the Philips Research Laboratories in Eindhoven, Holland and is currently in production at Magnavox Inc., the Philips production company based in the United States. Excellent descriptions of the M.C. 80 cryocooler were given by Haarhuis (1978) and by de Jonge (1979). A technical specification of the M.C. 80 was given by Haarhuis as follows:

Speed/Frequency: 3000 rpm/50 Hz
Input power: 125 VA (including 40 VA for air cooling fan)
Gas charge: helium at 16 bar
Compression ratio: 1.8
Refrigeration output: 1 watt at 80 K plus 0.035 W/K rise above 80 K
Minimum temperature with no load: 60 K
Ambient temperature range: -20 to 40°C
Cooling: Air
Weight: 7 kg
Vibration: < 30 μm
Noise level: < 55 dB
Attitude: not critical
Piston diameter: 16 mm
Piston stroke: 16 mm
Displacer diameter: 9.6 mm
Displacer stroke: 6 mm

From the diagrammatic cross-section of the M.C. 80 cryocooler shown in Figure 9.2 it will be noted that the piston and displacer arrangement are not the same as in the Beale unit discussed above. In the Haarhuis machine separate mechanical springs for the displacer and

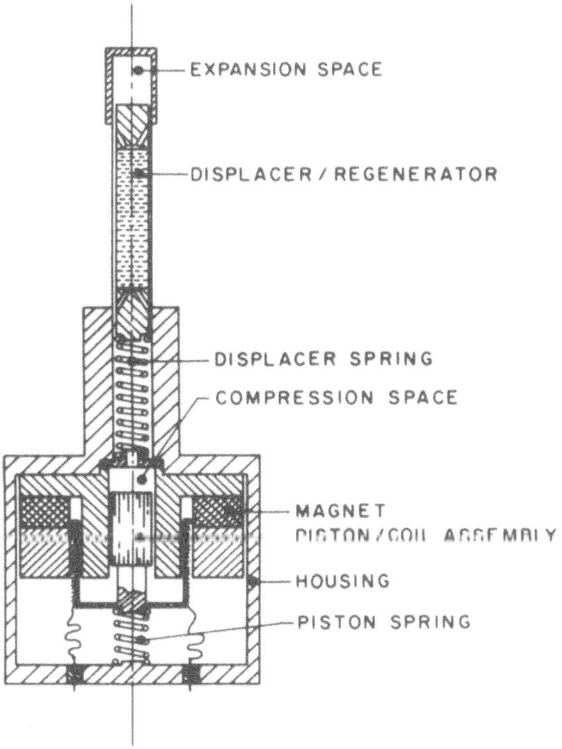

Figure 9.2 Cross-section of M.C. 80 integral free-piston Stirling cryocooler, (after Haarhuis (1978)).

the piston are provided rather than the gas springs of the Beale unit.

The M.C. 80 unit employs a permanent magnet with moving coil energized by an alternating current to drive the piston. The commercial versions utilize a standard magnet material Ferroxdure 300, also widely used for loudspeaker diaphragm drives. The military version, smaller, and with a lower power input requirement, uses Samarium/Cobalt magnets available at appreciably higher costs than the standard magnet materials.

SPLIT-STIRLING CRYOCOOLER

Split-Stirling cryocoolers are much favoured for infrared and electronic applications where there is a requirement for a low-mass gimbal mounted system or for minimum levels of mechanical vibration, or acoustic and electro-mechanical noise. Split-Stirling cryocoolers are in volume production at Cryogenic Technology Inc. (C.T.I.), Waltham, Mass., Martin-Marietta, Orlando, Florida and at the Hughes Aircraft Company, Torrance, California.

Figure 9.4 shows the Model CM-3 80 K cryocooler currently in

Figure 9.3 Model M.C. 80 integral free-piston Stirling cryocooler, (courtesy Philips Electrical Co.).

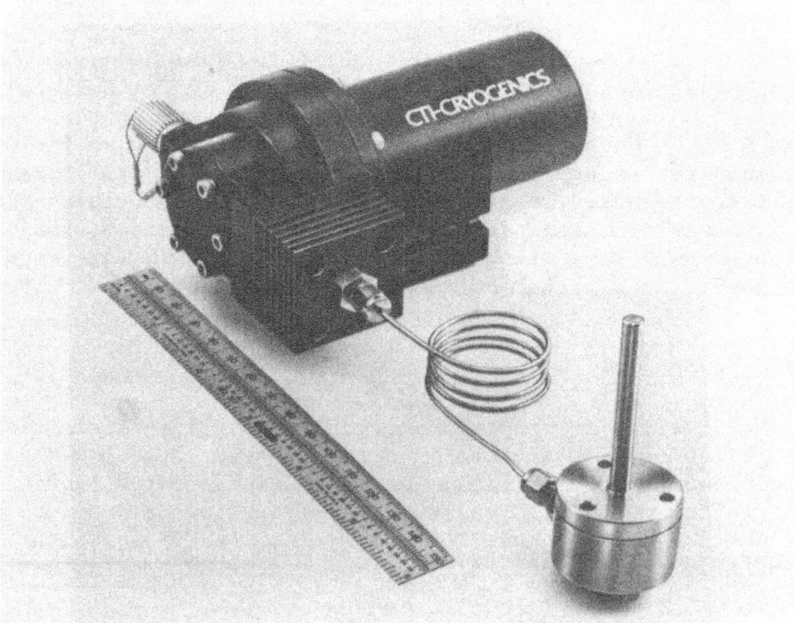

Figure 9.4 Model CM-3 split-Stirling cryocooler, (courtesy C.T.I. Inc.).

production at Cryogenic Technology Inc., Waltham, Mass. It consists of a compressor element and an expander element coupled by a single long tube. The compressor could be a free-piston device energized in some way to operate at resonant frequency with pneumatic or mechanical springs. However, in all the machines known to be in current production the compressor piston is activated by a crank/connecting rod mechanism driven by an electric motor. Nevertheless Ackermann (1981), presenting the dynamic analysis of a small free-piston resonant cryocooler, clearly had in mind a free-piston compressor driven by a linear electric motor similar to that discussed above in connection with integral Stirling cryocoolers.

The expander element of a split-Stirling cryocooler is shown in the diagrammatic cross-section given in Figure 9.5. The displacer

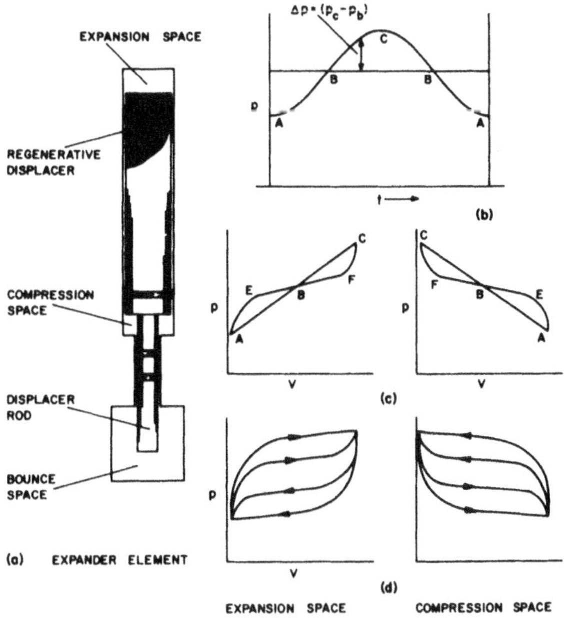

Figure 9.5 Expander element of split-Stirling cryocooler.

of the expander element could of course be coupled to and be driven by the same crankshaft as the compressor. However this would remove it from our special interest in free-piston (or free displacer) machines. It would also negivate the principal attraction of the split-Stirling cryocooler, the ability of the expander element to operate separate from and independent of the compressor. All machines of interest here are of the free displacer variety.

In most cases the units are small in size with a cold finger rarely exceeding 5.0 cm long and 6 mm in diameter. It is most convenient for the regenerator (fine copper or phosphur bronze screen, small metal spheres, etc.) to be incorporated into the displacer as shown in Figure 9.5.

A displacer rod, of substantial size is added to the displacer and extends through the seal element into the bounce space. The cold end of the expander cylinder is the expansion space. The ambient temperature end of the expander cylinder, below the displacer is part of the split compression space, the other part being above the piston and including the connecting duct.

When the compressor is operating the pressure of the working fluid will fluctuate periodically as shown in the pressure-time diagram, Figure 9.5b and the net fluidic force acting in the displacer is the difference in pressure between the working space and the bounce space times the displacer rod area. Let us assume first the pressure is the same in both the compression and expansion spaces and further that the pressure in the bounce space remains constant. Then if the displacer had no mass and was completely unrestrained in the cylinder it would move back and forth in the expander cylinder synchronously with the pressure difference between the working space and the bounce space.

The pressure-volume diagrams for the expansion and compression spaces would simply be the straight lines A-B-C shown in Figure 9.5c. The diagrams would have no area for the path of the upstroke diagram would coincide with that for the downstroke.

Of course the displacer does have a finite mass and so inertia forces would come into play to change the straight line linear diagrams A-B-C to the curved paths A-E-B-F-C. However as before the upstroke and downstroke diagrams would follow substantially the same paths so again no refrigeration would result from the displacer motion.

To achieve the objective of a significant regenerative effect it is therefore necessary to partially restrain the displacer at the start and throughout the motion both during the upstroke and also during the downstroke of the displacer. In this way the paths can be widely separated to produce the pressure volume diagrams for the compressor and expansion spaces shown in Figure 9.5d.

The partial restraint to displacer motion may arise from the frictional effects of the sliding seals provided at the bottom end of the displacer to separate the expansion and compression spaces so causing the working fluid to pass *through* the regenerator rather than the annulus between the displacer and cylinder wall. Another seal is provided on the displacer rod to separate the compression and bounce spaces. It is common practice for this latter seal to take the form of a long, very close tolerance, seal pair with the moving element, the displacer rod mounted to the displacer with the possibility of some free radial motion, to a limited extent, accommodating slight misalignments. The duty of restraining the displacer motion then falls principally on the lower displacer seal.

It is very difficult indeed to reproduce and maintain constant the seal friction restraint over the life of the unit. This is particularly true for the combination of helium working fluid necessary for cryocoolers and the presently available dry rubbing materials. Most success has been obtained with polytetrafluoroethylene (PTFE) based materials containing fillers of chipped glass or metal powders, i.e. Rulon 'A' etc. Technical exchanges among workers in the field (shop gossip) also favour the use of impure helium containing trace elements of others gases and vapours, principally water and hydrogen, but no scientific studies have been reported in the literature.

The ingenious, free displacer, split-Stirling system was first described by Dr. Walter Higa of the Jet Propulsion Laboratory, California, (Higa (1965)). It was conceived, independently, about the same time, by Mr. Fred Chellis of CTI, Waltham, Mass. and Dr. Stuart Horn of the U.S. Army Night Vision Laboratory, Fort Belvoir, Virginia.

Substantial efforts are being made to overcome the seal restraint problem. One solution proposed by Durenec (1977) is to use a small auxiliary compressor feeding the bounce space to provide a positive correctly phased driving force to the displacer. Other proposed solutions include the provision of a small electro-magnetic linear drive to the displacer (Ackerman (1982)).

Horn *et al* (1973) have presented an excellent theoretical analysis of the free displacer split-Stirling cryocooler but no detailed validation of the analysis by comparison with experimental data has been published in the unclassified literature.

LARGE STIRLING CRYOCOOLER

There are strong possibilities that future developments will include large Stirling refrigerators for gas liquefaction and relatively large scale refrigeration. Walker (1983) has discussed three Stirling machines of relatively large capacity that were developed and put into limited production. These were all machines with kinematic drive mechanisms. Other concepts for double-acting Siemens-Stirling engines are also under active consideration. These involve multiple cylinders interconnected so the expansion space of one cylinder is connected to the compression space in an adjacent cylinder. In this way the number of reciprocating elements can be halved to one per cylinder instead of the customary two per cylinder. Schemes have been postulated for multi-cylinder Siemens type free-piston engines with some expansion spaces heated to produce power while others operate as refrigerating machines. At least one version (Martini (1983)) has been reduced to practice.

Other schemes for large free-piston cryocoolers use the Stirling-Stirling or duplex-Stirling arrangement for gas-fired natural gas liquefier, heat pump or refrigeration applications. In this concept a Beale type free-piston Stirling engine with heated (by natural gas combustion or otherwise) expansion space produces work to drive a second Beale type free-piston engine acting as a refrigerator or heat pump. One arrangement of this system intended for use as a gas-fired natural gas liquefier is discussed in Chapter 8. Berkowitz (1982) has discussed a practical form of the engine development at Sunpower.

LIQUID-PISTON STIRLING CRYOCOOLER

Free-piston Stirling engines with liquid pistons are discussed in Chapter 10. The chapter was contributed by Dr. Colin West, inventor of the 'Fluidyne' liquid-piston Stirling engine and author of the definitive text on the topic (West (1982)).

It is possible the same principles set forth by West to operate a liquid-piston Stirling power system could also be utilized as a refrigerating system.

The conventional Fluidyne engine consists of a water (or other liquid) column in a 'U' tube and an associated 'tuning line'. The two limbs of the 'U' tube are coupled through a regenerative heat exchanger. One space above the liquid column acts as the compression space while the other acts as the expansion space. When the system is in proper proportion, and the expansion space is heated, the liquid columns oscillate with a phase difference of approximately 90° thereby creating a true Stirling engine. The pressure wave produced can be utilized to operate a water pump or some other work consuming device.

Now consider operation of the system as a refrigerator. In principle it appears feasible to drive a compressor remote from the cryocooler site to generate a pressure wave in a duct coupling the compressor and the liquid piston cryocooler.

Imposition of the pressure wave would cause the liquids in the 'U' tube to oscillate resulting in Stirling engine action with expansion of the working fluid in the expansion space and the generation of refrigeration if the space were unheated.

It is conceivable the liquid pistons could be the condensed phase of the system working fluid and, moreover, that, during expansion, part of the working fluid condensed and became part of the piston. Evaporation of some of the liquid constituting the piston is a common phenomenon in the 'wet' Fluidynes used for pumping.

If the condensed working fluid were the liquid of the piston it would of course be impossible to start up simply by operating the compressor. It would be necessary first to cool down the unit to its equilibrium, operating temperature and 'prime' the unit with condensed working fluid to provide the 'piston' with which to start. Thereafter operation of the compressor could cause the system to generate sufficient refrigeration to maintain stable operation.

Successful development of this system could provide a cryocooler of virtually infinite life apart from periodic replacement of the compressor as necessary. Long lived cryocoolers and refrigerators are sought for every application but are of particular importance for the superconducting applications foreseen for the near future, the Josephson tunnel diode 'shoebox' computer, the many applications of 'SQUIDS', superconducting quantum interference devices, the superconducting magnets for magnetohydrodynamic power generation, ship propulsion and the like. For all these applications a long life liquid helium 4 K cryocooler is highly desirable.

CONCLUSION

Cryocoolers for a variety of military and civil applications are the only well established application of Stirling engines at the present time. As we enter the 'coming age of superconductivity' opportunities for further applications of Stirling cryocoolers will dramatically increase.

Free-piston Stirling cryocoolers of both the integral-Stirling and split-Stirling variants are presently in quantity production and serious consideration is being given to larger units for a variety of applications involving moderate scale gas liquefaction and refrigeration. Thermally activated (gas-fired) cryocoolers, refrigerators and heat pumps become increasingly attractive as natural gas emerges as the preferred industrial commercial and domestic primary fuel.

REFERENCES

Ackerman, R. (1981). Dynamic Analysis of a Small Free-Piston Resonant Cryorefrigerator. pp. 57-69, Refrigeration for Cryogenic Sensors and Electronic Systems, N.B.S. Special Publication 607, Eds. Zimmerman, J.E., Sullivan, D.B. and McCarthy, S.E., U.S. Govt. Printing Office, Washington.

Ackerman, R. (1982). Contribution to Conference on Small Cryocoolers for Electronic Applications. NASA Goddard Spaceflight Center, Greenbelt, Maryland, November, (Max Gasser).

Berkowitz, D. (1982). The Design, Development and Performance of a Duplex Stirling Natural Gas Liquefier. Paper No. 829296, Proc. 17th Inter. Soc. Ener. Conv. Eng. Conf., Los Angeles, August.

de Jonge, A.K. (1979). A Small Free-Piston Stirling Refrigerator. Paper No. 799245, Proc. 14th Inter. Soc. Ener. Conv. Eng. Conf., Boston, Mass., pp. 1136-1141, August.

Durenec, P. (1977). Private Communication. (see also Walker, G. (1977). Split-Stirling Cooling Engine with Fluidic Control of Displacer Motion. Invention Disclosure, University of Calgary, Alberta, Canada, Dec.).

Haarhuis, G.J. (1978). The MC80-A Magnetically Driven Stirling Refrigerator. Proc. 7th Int. Cryo. Eng. Conf., London, IPC Business Press, Guildford, U.K.

Higa, W. (1965). A Practical Philips Cycle for Low Temperature Refrigeration. Cryogenic Technology, Vol. 8, pp. 203-209, July/August.

Martini, W. (1983). Private Communication. Martini Engineering Inc., 2303 Harris, Richland, Washington.

Walker, G. (1983). Cryocoolers. 2 Vols., Intl. Mono. on Cryogenics, Plenum Press, New York.

Walters, B.T. and Acrod, T.T. (1973). Miniature Cryogenic Cooler for TOW Night Sight. Proc. Closed Cycle Cryocooler Tech. and Applics., Vol. 1, pp. 55-72, AFFDL-TR-73-149 WPAFB, Ohio, AD, No. 918234.

West, C.D. (1982). Liquid Piston Stirling Engines. Van Nostrand Reinhold Co. Inc., N.Y.

CHAPTER 10 - LIQUID PISTON STIRLING ENGINES

Dr. Colin West, Box 262A, Oliver Springs, Tenn. 37840, contributed this chapter on liquid piston Stirling engines - it gives the flavour of his recent more extensive work of the same title. Colin West worked at the British Government research establishment, the Atomic Energy Research Establishment, at Harwell under the supervision of Mr. E.H. Cooke-Yarborough (see Chapter 6) throughout the 1970's. Among other projects he worked on the diaphragm Stirling engine described in Chapter 6, the Harwell TMG. It was during this time that Colin invented the liquid piston Stirling engine that has remained a prime interest source.*

He now works at the U.S. Government Oak Ridge National Laboratory, Tennessee, and continues developmental studies of liquid piston Stirling engines in his spare time.

G. Walker

The Fluidyne liquid piston engine is a simple free-piston Stirling engine that can be made from nothing more than some lengths of tubing. The pistons are liquid, so they always fit the cylinder exactly with no need for machining, close tolerances, mechanical seals or bearings. On the other hand, the liquid piston engine in its simplest form operates at close to atmospheric pressure, and at a low frequency - typically 1/2 to 1 Hz. Consequently, the power density is low, and a large machine is needed if a substantial power output is required. These factors, along with the convenience with which a liquid piston machine can be adapted to pumping, have defined the main fields of potential application. Most interest so far has centered around the use of Fluidynes to pump water, particularly for irrigation or drainage pumping in developing countries or in more specialized circumstances where electric power may not be reliably available. However, as any other Stirling machine, the liquid piston engine can be operated as a refrigerator or heat pump and several workers have proposed exploiting this. Walker has suggested that a liquid helium cooler - for example, to service superconducting computers - could be made in which the liquid helium would be contained in a Fluidyne machine, driven by gas pressure variations from outside the cryogenic region. W. Martini has demonstrated that the liquid piston engine can be operated as a heat-actuated heat pump.

In this chapter we shall discuss some of the characteristics of the liquid piston engine, and draw attention to some of the effects observed in these machines that are not present or are less important

---

**Liquid Piston Stirling Engines*, 1982.

in other free-piston Stirling engines.

BASIC PRINCIPLES

The basic principle of the liquid piston Stirling engine, known as the Fluidyne, is shown in Figure 10.1. In this example, both the displacer and the piston are liquid. If the displacer column is

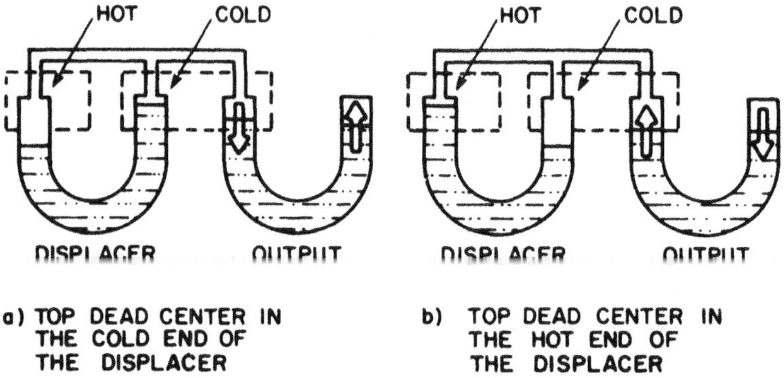

Figure 10.1 Basic operation of the Fluidyne.

set into oscillation in its U-tube, the gas above the liquid surface is transferred back and forth between the hot and cold spaces. The resulting pressure variation acts on the liquid in the output column, causing it to move also. The pressure variations do not, however, have any effect on the displacer for they act equally on both ends of the displacer column. If the displacer were left to itself, therefore, the oscillations in the displacer tube would eventually die away because of viscous friction and other losses. Several means are known for feeding back energy from the power piston to the displacer in order to overcome these losses; they are discussed below.

TUNING OF LIQUID COLUMNS

The maximum amplitude of movement in the output column will occur when the driving force has a frequency almost equal to the resonant frequency of the water oscillating in the output column (Chapter 2). The driving force in this case is the pressure variation due to the oscillations of the displacer liquid, and so it follows that for maximum movement the natural or resonant frequencies of the two columns should be approximately equal.

What happens if, say, the water column length in the output tube is not adjusted to give it the same natural frequency as the displacer? If the output column is too long, the mass of water in it will be so great that the pressure change will be unable to move it very far, and there will be almost no change in the volume of the working gas during the cycle; therefore, the indicated power will be very small. On the other hand, if the water column is too short, it will move so easily that the gas pressure will be unable to build up significantly before the column moves to its full extent, and again the indicated power will be very small. For maximum output, therefore, the length of the output U-tube must be "tuned" to suit the operating frequency of the engine; the load on the output column may also need to be taken into account in calculating the optimum length.

Usually, the natural frequency of the displacer will determine the operating frequency of the engine because it has relatively lower losses than the output column, being both shorter and wider as we shall see. In the case of a simple U-tube displacer (Figure 10.2), calculating the natural frequency of the liquid oscillating under the restoring force of gravity is a standard text-book problem like the oscillation of a pendulum:

$$f = \frac{1}{2\pi}\sqrt{\frac{2g}{L_d}} \qquad (10.1)$$

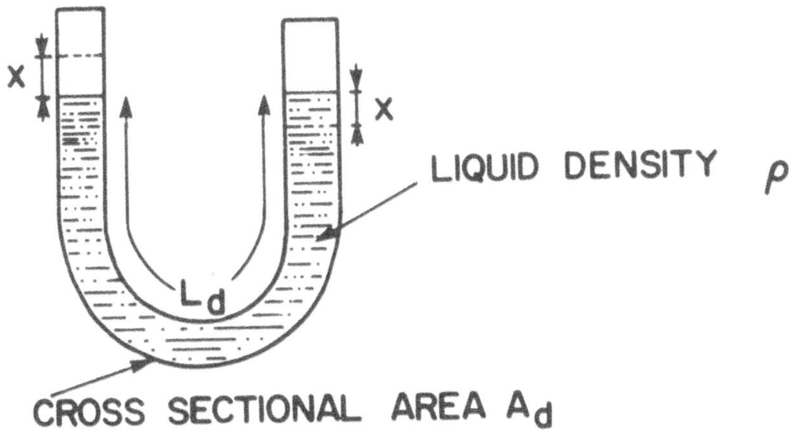

Figure 10.2 Simple displacer U-tube.

Sometimes, it is convenient to connect the hot and cold displacer cylinders through a common reservoir rather than making them from a single U-tube (Figure 10.3). In this case, the natural frequency is given by:

$$f = \frac{1}{2\pi}\sqrt{\frac{g(1/A_1 + 1/A_2)}{h_1 + h_2}} \qquad (10.2)$$

This equation, and several others relating to freely oscillating

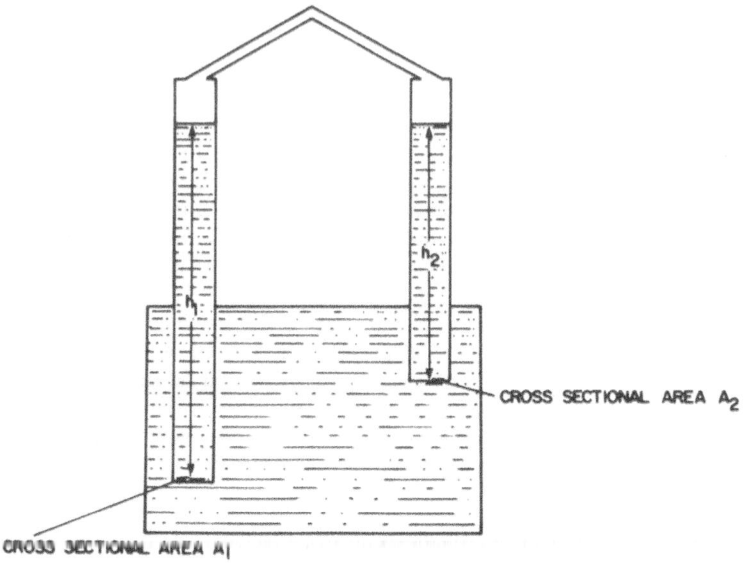

Figure 10.3 Reservoir displacer.

liquid columns, are derived in the book 'Liquid Piston Stirling Engines', (West (1982a)).

The calculation of a natural frequency for the output column is complicated by the fact that movement of this liquid will compress or expand the working gas, thus giving rise to an extra restoring force in addition to gravity. For the configuration shown in Figure 10.4, the extra pressure ΔP due to a movement x of the output column (often called the tuning line) is approximately given by:

$$\Delta P \simeq \frac{P_m A_t x}{V_m} \tag{10.3}$$

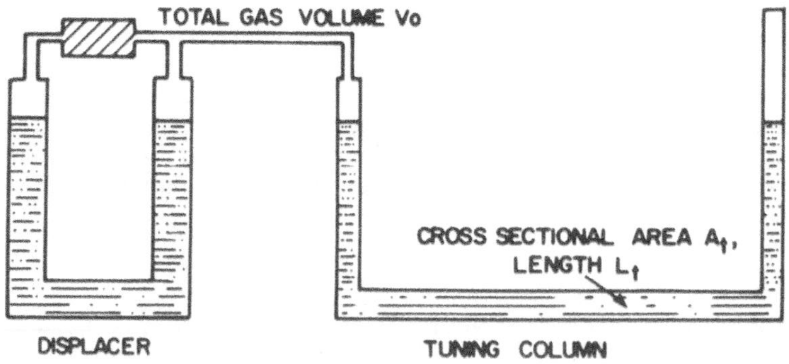

Figure 10.4 Tuning column configuration with separate cylinders.

This must be added to the pressure due to the difference in height, 2x, of the ends of the liquid column. The total pressure acting to return the liquid to its equilibrium position is therefore:

$$\Delta P \approx \frac{P_m A_t x}{V_m} + 2\rho g x \tag{10.4}$$

The mass of liquid in the tuning column is $\rho A_t L_t$, and so its natural frequency is given by:

$$f = \frac{1}{2\pi}\sqrt{\frac{2g}{L_t} + \frac{A_t P_m}{\rho V_m L_t}} \tag{10.5}$$

This equation is modified somewhat if some of the gas spaces are adiabatic rather than isothermal, or if a merged configuration is adopted (Figure 10.5), but the principle is unchanged.

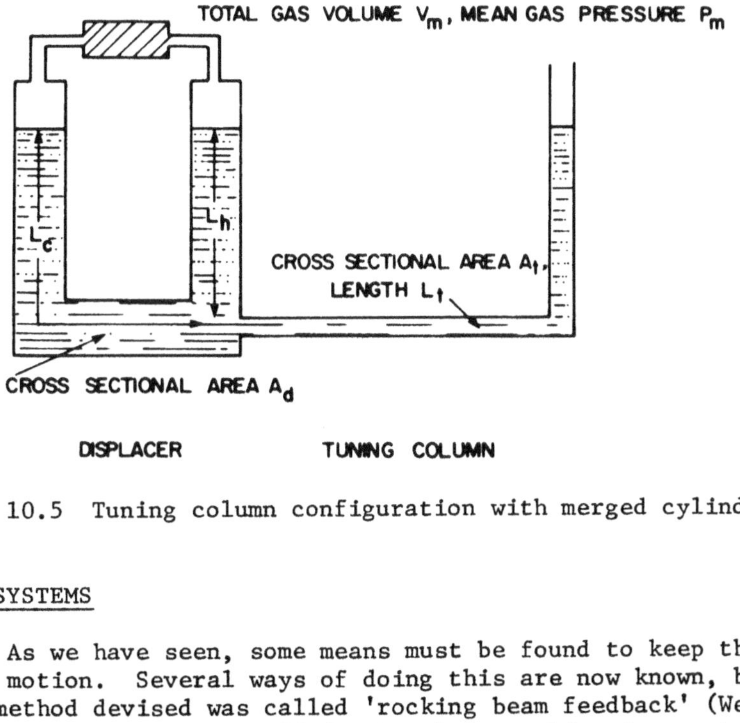

Figure 10.5 Tuning column configuration with merged cylinders.

FEEDBACK SYSTEMS

As we have seen, some means must be found to keep the displacer in motion. Several ways of doing this are now known, but the earliest method devised was called 'rocking beam feedback' (West (1970)). As shown in Figure 10.6, the whole machine is mounted on a pivot, flexure or spring, and as the liquid in the output tube moves back and forth, its shifting weight causes the whole machine to rock like a seesaw. This rocking movement keeps the displacer liquid in motion. A spring may be used to provide some extra restoring force and to adjust the frequency of the rocking motion to match

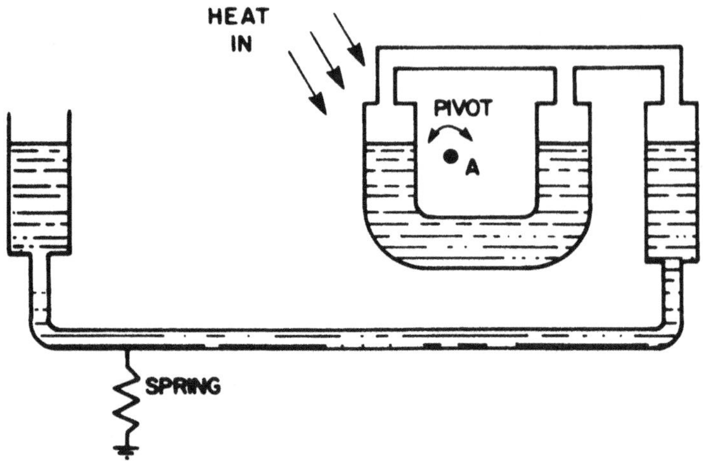

Figure 10.6 Rocking beam feedback for a displacer Fluidyne engine.

approximately the other frequencies in the system.

Another type of rocking beam feedback (West (1974a)) is shown in Figure 10.7. In this case, the rocking motion is driven by the action of the pressure variations in the working fluid acting on a bellows or other flexible coupling.

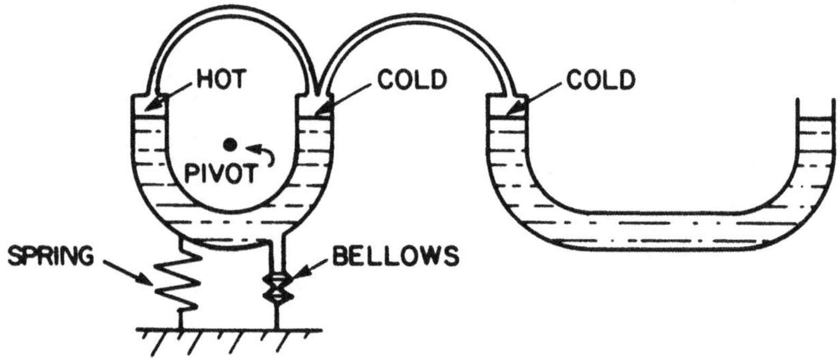

Figure 10.7 Rocking beam engine driven by pressure variations in the working field.

The rocking beam system can also be adapted to the alpha configuration Stirling engine (Figure 10.8) thus decreasing the number of cylinders to two and reducing the unswept volume. However, in doing this the possibility is lost of using two different liquids in the displacer and output tube, not in contact with each other, unless some kind of separating membrane is used.

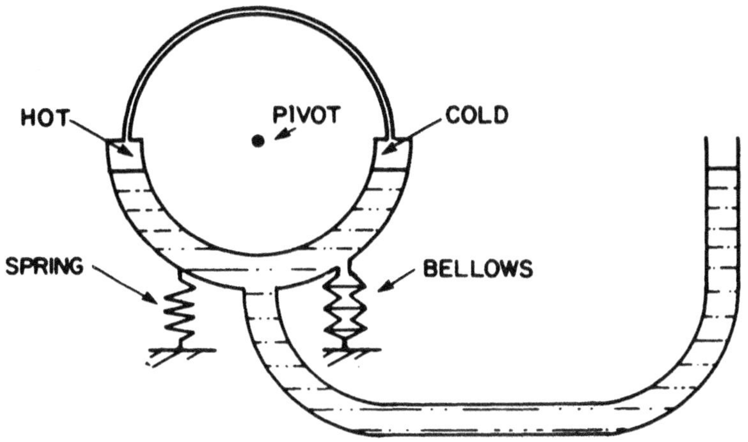

Figure 10.8 Pressure driven feedback for a rocking beam machine in the merged cylinder of alpha configuration.

Most work to date has concentrated on Fluidynes without mechanical moving parts, in which the feedback is provided by the motion of the liquid itself. One example of this, first proposed by Cooke-Yarborough, and which has been widely adopted because it is flexible and fairly insensitive to mistuning, is shown in Figure 10.9. To explain the operation of this system, we begin by recalling

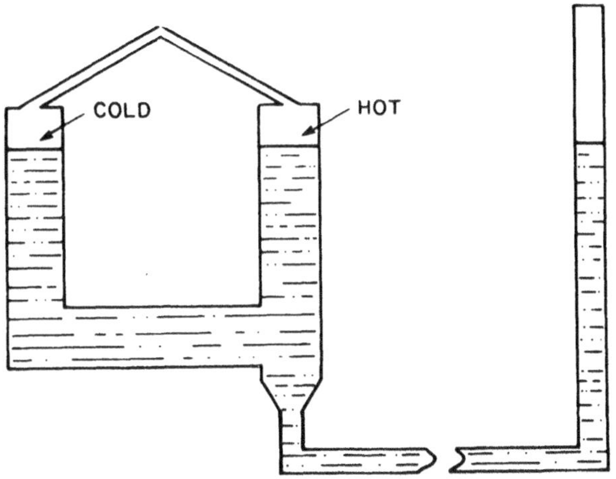

Figure 10.9 Liquid feedback machine (tuning line pointing toward hot end).

that to operate as an engine, the movement of the hot piston must lead the movement of the cold piston - that is, the hot piston should reach bottom dead center, for example, somewhat earlier in the cycle than the cold piston. Now, in the engine illustrated in Figure 10.9, the length of the hot column - measured from the free surface to the junction with the output tube - is less than the length of the cold column. Consequently, there is less mass of liquid in the hot side of the engine than in the cold, and when the pressure varies in the working fluid, the hot column responds more quickly than the cold, i.e. the hot cylinder volume variation leads the cold.

This argument, which is rather oversimplified, implies that the machine can run as an engine regardless of the orientation of the output tube (see for example, Figure 10.10), as long as the junction is

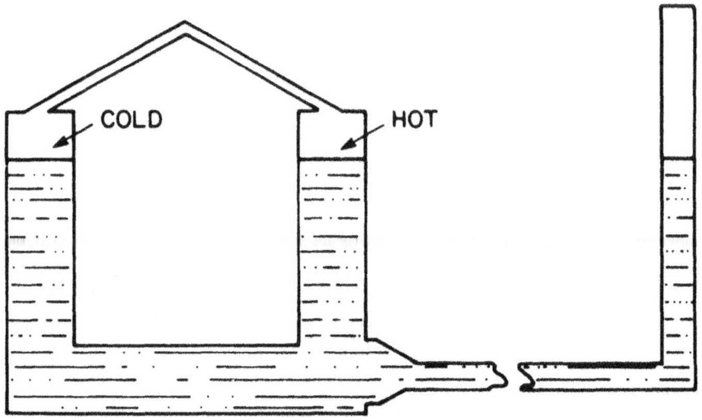

Figure 10.10 Liquid feedback machine (tuning line pointing toward cold end).

closer to the heated than to the cooled cylinder. A somewhat more rigorous explanation (West (1983a)) bears out this conclusion, which is consistent with observations made on Fluidynes operating on an air cycle. However, Goldberg (1979) reported the operation of a machine in which water was allowed to evaporate in the hot cylinder, so that the working fluid, an air/steam mixture, was dependent on the orientation rather than the position of the junction: he noted that the operation of this machine could be achieved with the junction close to either the hot or cold cylinder - but only if the end of the output tube was directed toward whichever working space (hot or cold) was the closer. At present, there is no explanation of these differences in observed behavior.

The liquid feedback system is adaptable to many different applications of the basic machine (West (1971)). Figure 10.11 shows a simple arrangement with the displacer column divided into two unequal lengths communicating with each other, and with the output tube. Figure 10.12 shows an arrangement (West, Geisow and Pandey (1976)) in which the hot and cold pistons are concentric cylinders. This configuration has been used to build a very large machine with a throughput of up to 4000 gals. per hour (Pandey (1981a)).

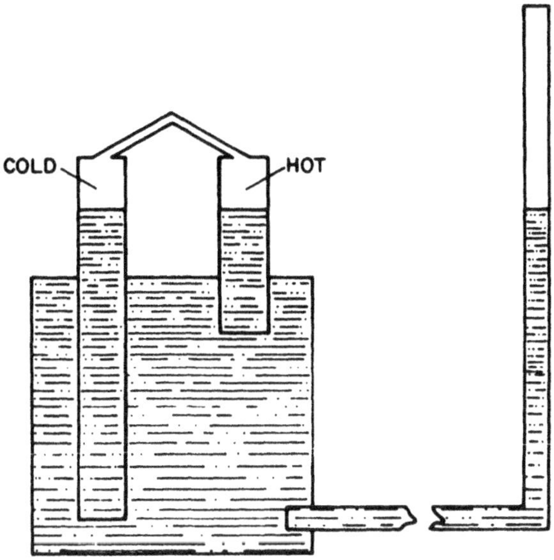

Figure 10.11 Reservoir and unequal cylinder lengths machine.

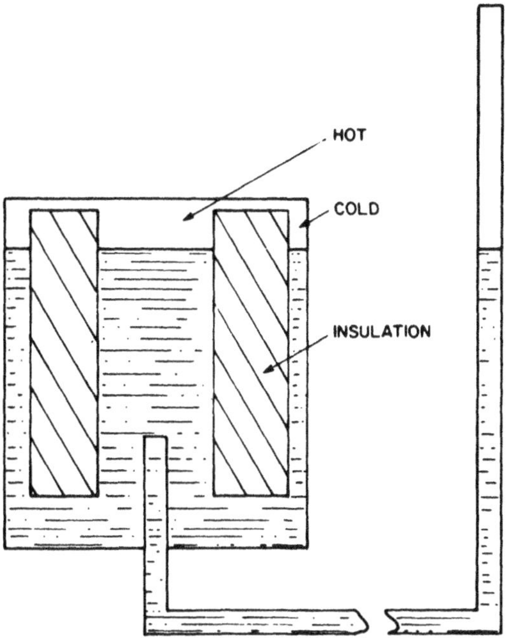

Figure 10.12 Concentric cylinder machine.

Elrod's analysis (Elrod (1974)) describes the mathematical behavior of liquid feedback machines using a linearized approximation which will be valid for small amplitudes and neglecting losses and load terms. Geisow's published analysis (Geisow (1976)) used the same mathematical model, but in unpublished work he extended the model to include the effects of some damping terms. The effect of losses and loading is included in Stammers analysis of the problem (Stammers (1979)); this analysis is very important in explaining and predicting the behavior of real machines operating at large amplitudes. A more recent discussion (West (1983a)) uses the method of vector analysis applied to the linearized differential equations of motion to present a simple physical explanation of the liquid feedback system.

A number of computer analyses have been carried out (Gosling and Boast (1976), Lewis (1978), Drzewiecki (1979), Singleton (1979), Gill (1980)) based on various mathematical models. Among these, the work of Gosling and Boast was unusual in that it was carried out with an analog computer; this proved to be a powerful and entertaining tool for the exploration of the Elrod model.

In the multicylinder Fluidyne shown in Figure 10.13 (West

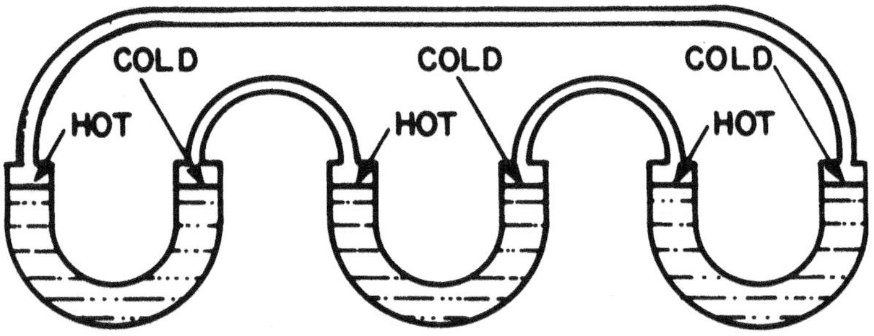

Figure 10.13 Multicylinder machine.

(1974b)), each column acts as both displacer and output line, as in a Siemens double-acting engine. In this case, each column is subject to gas pressure forces as well as to gravity, and no separate feedback system is needed. A similar arrangement was proposed independently by Finkelstein, and was the subject of some experimental measurements (Cutler and Hanke (1979)). The concentric cylinder arrangement can also be used for multicylinder machines.

Most interest in the multicylinder configuration appears to be centered around the possibility of configuring the system as a heat-actuated heat pump (West, Geisow and Pandey (1976), Gerstmann and Friedmann (1977)), and this concept has now been experimentally verified (Martini (1983)).

PUMPING CONFIGURATIONS

So far, we have discussed several forms of the liquid piston engine but little about how the engine output power is to be extracted and used. The most obvious application for the Fluidyne - besides refrigeration and heat pumping - is in pumping, including irrigation pumping and drainage.

The direct mechanical output from a rocking beam machine may be used with a simple valveless pump (Figure 10.14) that can, if desired, be mounted on the same pivot as the beam. Of course, the rocking motion could also be used to turn a crankshaft or move a slider, and thus operate a conventional mechanical pump.

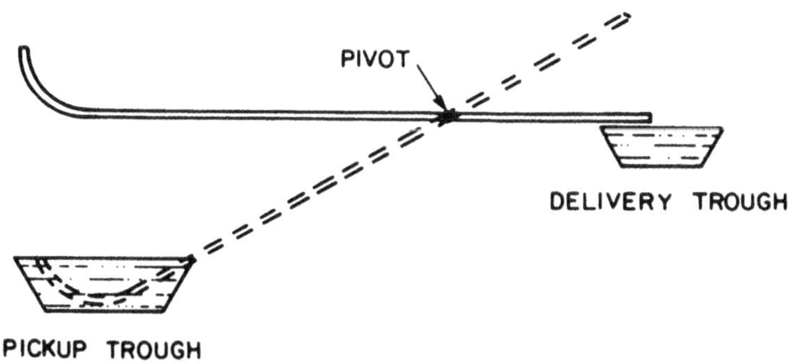

Figure 10.14 Valveless pump for a rocking beam machine.

Besides these direct mechanical systems, there are three simple ways to use the Fluidyne output to pump water, or another liquid.

The first, known as 'series coupling', simply requires a T-piece connection at the end of the output tube and two non-return valves (Figure 10.15). Most small working models of Fluidyne pumps have used this method.

The second method (Figure 10.16) places the pumping system closer to the displacer end of the output tube, so that it is effectively in parallel with the output column. In this case, the volume of liquid moving in the output tube can be much greater than the volume passing through the pump; consequently, the pumping head and throughput and nonlinearities in the pumping system, have less effect on the engine tuning. The output column does no direct work, except to overcome its own losses, but merely oscillates at a frequency tuned to that of the displacer, thus giving rise to a relatively large pressure oscillation in the engine for pumping. For this reason, the output tube is often referred to as the 'tuning line' or 'tuning column', recognizing that its main function is to have a large, resonant oscillation and not

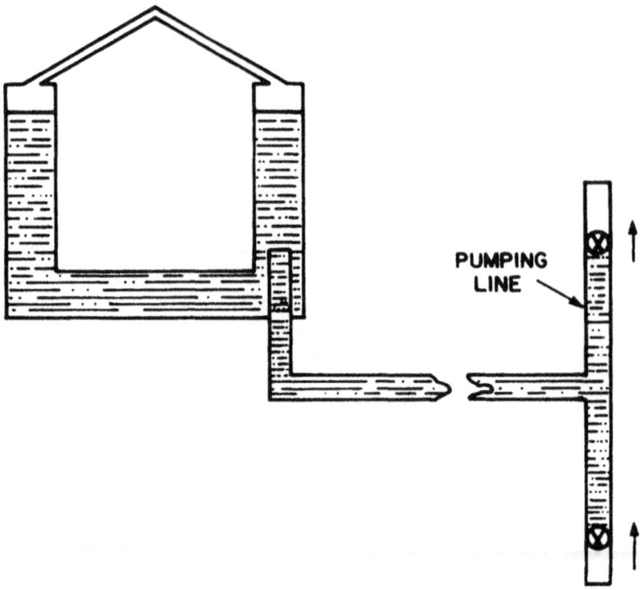

Figure 10.15 Pump in series with output column.

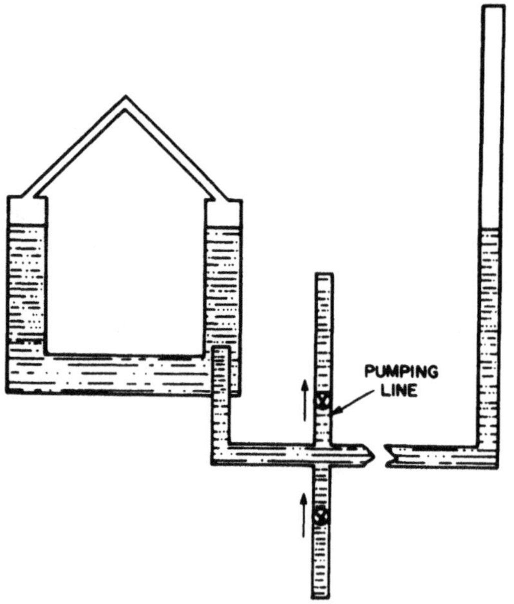

Figure 10.16 Pump in parallel with tuning column.

to provide a direct output mechanism.

In the third method, the pump is not connected to the tuning

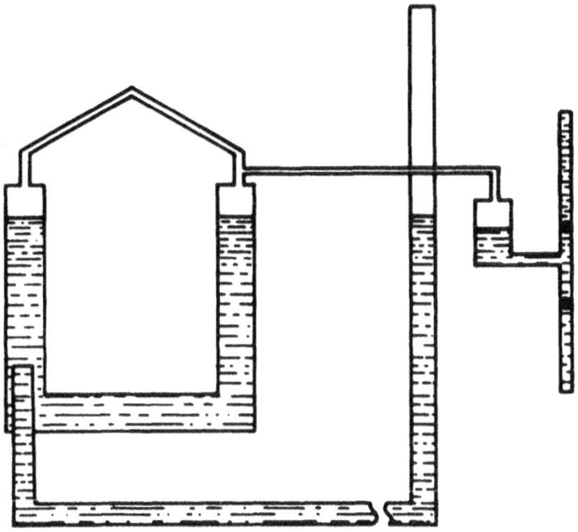

Figure 10.17 Pump gas coupled to cold space.

column at all (Figure 10.17), but is driven instead from the pressure variations in the working gas. This method of connecting the pump, known as 'gas coupling', can in principle, be applied to the rocking beam, liquid feedback and multicylinder machines alike. It has the important advantage, first proposed by Cooke-Yarborough, that the pump and engine need not be at the same level.

Both series- and gas-coupled pumping systems have been made at Harwell using fluidic valves connected to liquid feedback Fluidynes. In this way, a simple heat actuated water pump can be made, having no moving parts, and hence no wear.

WET AND DRY MACHINES

Most of the experimental machines described in the open literature have permitted substantial evaporation to take place from the liquid in the hot cylinder. The most obvious result is an increase in the heat input required, since latent heat must be supplied, but a second effect is to increase the power output. The enhancement of output occurs because the pressure variations caused by evaporating and condensing water at the hot and cold ends are greater than would be caused by displacing air between the same temperature limits. Calculations show that if the evaporation and condensation are restricted to the most advantageous parts of the cycle, a five or six fold increase in indicated power output may be achieved (West (1983b)).

Unless precautions are taken, or unless the vapor pressure of the liquid in the hot cylinder is much less than the mean pressure of the working gas, evaporation will take place in a rather uncontrolled way. Machines of this kind are called 'wet Fluidynes'. By using a low vapor pressure liquid or low temperatures, or by placing a float on top of the liquid in the hot cylinder to insulate it from high temperatures, evaporation can be suppressed or made negligible. Machines of this type are called 'dry Fluidynes'. The solid displacer, liquid output piston machine (Figure 10.18) is also a 'dry' engine in this sense.

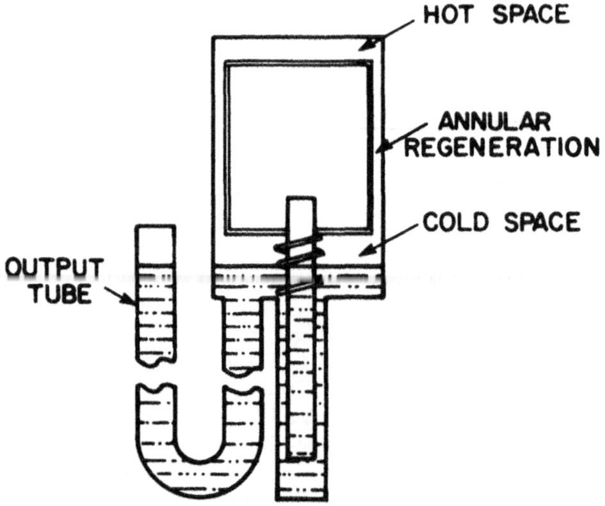

Figure 10.18 Solid displacer, liquid piston engine, (after West, Cooke-Yarborough and Geisow (1970)).

Table 10.1
Operating Regimes for Atmospheric Pressure Fluctuations

Type of Machine	Hot Space Temperature	Cold Space Temperature	Comments
Wet	370K	300K	Air cycle is dominant.
Wet	470K	320K	Water vapor cycle and closed regenerative air cycle are simultaneously evident.
Dry	500K	350K	Closed regenerative cycle using dry air.

Reader and Lewis (Reader and Lewis (1979b)) identified the three operating regimes revealed in their experiments and described in Table 10.1. The overall efficiency may be limited by the extra heat needed to support evaporation, although in small machines especially, other losses may be so large that the overall effect of evaporation is to increase the efficiency from a negligible or zero value to a small value. In any event, the maximum efficiency reported for a wet machine is less than 1/2 percent, whereas a figure of 7 percent has been quoted for a large dry machine. More details of this are given in a paper comparing the performance of wet and dry Fluidyne engines (West (1982b)).

FLOW EFFECTS

Most of the fluid flow problems met by engineers concern steady flow in long ducts. In a Fluidyne, however, the flow is oscillating, and this modifies the flow behavior profoundly, as first pointed out by Elrod.

The problem of oscillating flow was investigated in the last century (Rayleigh (1896)) by Kirchoff and other physicists studying the behavior of sound waves in tubes or porous material. An analysis of oscillating flow in the Fluidyne was carried out by Ryden (1975) but his work remains unpublished. A useful and convenient formulation of the problem, applicable to the Fluidyne system, has been given by Crandall (1927).

The laminar behavior of a fluid oscillating in a tube depends on the diameter of the tube. If the tube is "narrow", the flow pattern is well described by the usual Poiseuille laws and the velocity profile is parabolic. If the tube is "wide", the flow effects are limited to a relatively thin boundary layer close to the wall, with the fluid in the center of the tube moving as if it were solid (see Figure 10.19).

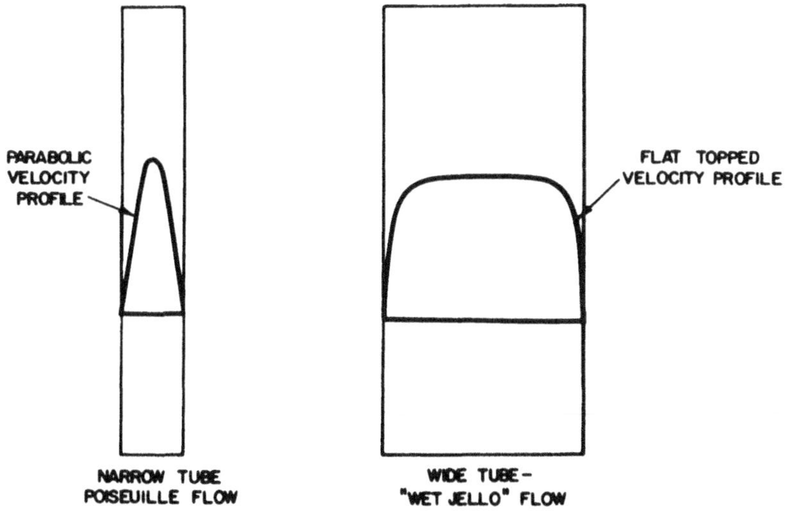

Figure 10.19 Velocity profiles for oscillating flow in narrow and wide tubes.

According to Crandall, the thickness of this boundary layer, t_d, is a function of the liquid density and viscosity and the frequency of oscillation. As long at the flow is laminar, it is not dependent on the amplitude

$$t_d = \sqrt{(\eta/\pi\rho f)} \qquad (10.6)$$

If the tube diameter is greater than about three times the boundary layer thickness, it may be considered "wide". If the tube diameter is less than half this, it may be considered "narrow". Intermediate sizes show a more complex flow pattern than either extreme.

For water at room temperature, oscillating at say 0.5-1 Hz (typical of Fluidyne), the boundary layer thickness according to Equation (10.6) is less than 1 mm, and tubes with a radius greater than a few millimeters will show the modified flow behavior. In air, the boundary layer thickness will be 3-5 mm, and tubes with a radius less than about 5 mm will obey the Poiseuille laws.

One important difference between the narrow and wide tubes is in the flow resistance, defined as the pressure drop per unit length divided by the mean flow velocity. Another important effect is that in the wide tubes, the Reynolds number for the transition to turbulent flow may be very much higher than for Poiseuille flow (Park and Baird (1970)). These factors are illustrated in Table 10.2.

Table 10.2
Oscillating Flow Effects in Tubes

Radius[a]	$\leq 4.5 \sqrt{\omega\rho/\eta}$	$\geq 8.5 \sqrt{\omega\rho/\eta}$
Resistance Coefficient	$8\eta/R^2$	$\sqrt{2\rho\omega\eta}/R$
Critical Reynolds Number[b]	2,000	$375(R^2 \omega\rho/\eta)^{2.3}$

[a] According to Chan and Baird (1974).

[b] Park and Baird formula (1970) modified to take account of their experimental finding that their theoretical formula underestimates the critical Reynolds number by 50%.

Note that the resistance coefficient is independent of frequency in Poiseuille flow but increases as the square root of frequency in the wide tube case. Furthermore, in the wide tube case the resistance coefficient does not decrease as rapidly with increasing diameter as it does in Poiseuille flow. However, it increases only with the square root of the viscosity.

A secondary effect of the modified flow behavior is that

even with liquid pistons there are significant shuttle, or step down, losses. We have seen that the liquid in a wide tube oscillates as though it had a solid core surrounded by a boundary layer, or gap whose thickness t_d, is given by Equation (10.6). The shuttle losses for a liquid piston in these circumstances can be calculated, at least approximately, by treating it as a solid piston separated from the cylinder wall by a gap t_d filled with a fluid having the bulk properties of the liquid. The losses may then be calculated as for a solid piston (see Martini (1978)). The shuttle losses for an oscillating liquid can be very high, especially for water because of its relatively high conductivity and low viscosity (a low viscosity leads to a thin boundary layer and hence a narrow effective gap). An oil would show lower shuttle losses and higher viscous flow losses.

TRANSIENT HEAT TRANSFER LOSS

As the gas in the machine is expanded and compressed, its temperature tends to fall and rise. However, the gas temperature immediately adjacent to any solid surface (such as the cylinder walls) hardly changes at all. There is, therefore, a continually varying temperature gradient between the bulk of the gas and the region adjacent to the walls. Heat conduction down this temperature gradient is an irreversible process leading to a loss of efficiency. This effect takes place in gas springs and in the cylinders of an engine.

The presently published analyses of the effect (Breckenridge, Heuchling and Moore (1971), Lee, Smith and Faulkner (1980)) use a model that takes into account only conduction in an undistributed gas. In practice, mixing of the gas, and scraping of the boundary layer by the moving piston or by the incoming gas, enhance the heat transfer effects and lead to greater losses: this is accounted for by an experimentally determined enhancement factor. Breckenridge *et al* (1971) reports an experimentally determined factor of 2.6 for the gas spring that they tested. Sunpower Inc. (Wood (1980)) reports an enhancement factor of 5 to 8, with the lower values corresponding to gas spring experiments and the higher values to engines. Lee has reported enhancement factors in the range of 5-10 are typical of Stirling engines. The actual value probably depends on frequency, gas filling, cylinder geometry and other variables:

$$\text{Loss per cycle} = \frac{F\times\pi}{\sqrt{2}} \left(\frac{\gamma}{\gamma-1}\right) A_s\, P_m \left(\frac{\Delta T}{T_m}\right)^2 \sqrt{\alpha/\omega} \tag{10.7}$$

$$= F\,\sqrt{\pi/2}\,\left(\frac{\gamma}{\gamma-1}\right)^{1/2} A_s \left(\frac{\Delta T}{T_m}\right)^2 \sqrt{kT_m}\sqrt{\frac{P_m}{f}} \tag{10.8}$$

where
F = enhancement factor,
f = frequency,
α = thermal diffusivity of the gas,
γ = gas specific heat ratio,

A_s = surface area,
P_m = mean pressure,
ΔT = amplitude of temperature variations,
T_m = mean temperature,
ω = angular frequency,
k = thermal conductivity.

From the above it is clear the loss per cycle increases as the square root of the pressure, and decreases as the square root of the frequency. The gross output per cycle of an ideal machine is directly proportional to the mean pressure and independent of the frequency. Therefore, the losses are a smaller proportion of the gross output at high pressures and frequencies.

Equation (10.8) implies that the loss per cycle becomes indefinitely large as the frequency becomes lower and lower. This will not happen, of course, for the reason that at very low frequencies the cylinder behaves isothermally and there are no temperature variations, i.e. ΔT tends to zero. This is illustrated in Figure 10.20. For any particular combination of machine size, gas filling, temperature and pressure, there is likely to be a frequency at which the losses are highest. Raising or lowering the frequency from this value would reduce the loss per cycle.

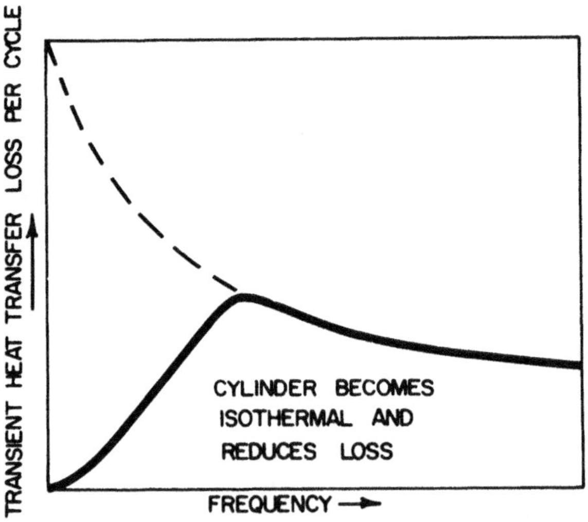

Figure 10.20 Transient heat transfer loss as a function of frequency.

RESULTS

Many groups and individuals have built and operated Fluidynes and reported their observations. Many of these observations are descriptive or qualitative, but we summarize here some of the quantitative performance data reported. Most work has been done with wet

Fluidynes (in many ways, easier to build and operate), but exceptions include Martini, Hauser and Martini (1977), Goldberg and Rallis (1979), West and Pandey (1981) and Pandey (1981b).

The published performance results are summarized in Table 10.3. The table gives data on all three of the most important performance parameters (throughput, head and efficiency) where available.

Table 10.3
Published Performance Figures for Fluidyne Pumps

	Performance		
	Flow Rate* (U.S.gph)	Head* (feet)	Efficiency* (percent)
West (1970)	3	3.3	-
West (1971)	100	5.3	0.35
Goldberg *et al* (1977)	9.5	2.0	0.12
Goldberg *et al* (1977)	11.5	3.0	0.08
Mosby (1978)	5.9	1.0	0.15
Reader (1979)	-	-	0.03
Bell (1979)	30	3.6	0.18
West and Pandey (1981)	460	10.0	4.7
Pandey (1981a)	2500	10.0	7.0

*Figures quoted are fairly typical, but were not necessarily achieved simultaneously.

There has been little work published on multicylinder configurations. The only known experimental results were obtained by researchers at the Chicago Bridge and Iron Company (Cutler and Hanke (1979)) who tested two machines. This group, working with T. Finkelstein, have apparently made important advances in balanced compounded Stirling hydraulic machines but no details have been released. Martini (1983) has reported the experimental observation of cooling in a device that may be described as a heat-powered, multicylinder liquid piston refrigerator or heat pump.

ENGINES FOR BEGINNERS

For newcomers to the field, two drawings of actual small scale working Fluidynes are included here (Figures 10.21 and 10.22). The first is a wooden engine of simple construction heated by a resistor inside the hot cavity. Thin aluminum foil is folded into four or five configurations and placed in the cold cylinder to act as an isothermalizer. Extra cooling for the water is provided by the external cooling water reservoir: as the pressure in the engine rises, some warm water is forced from the cold cylinder into the reservoir, by compressing the air above the water surface there, and mixes with the much larger quantity of cooler water already in the reservoir. As

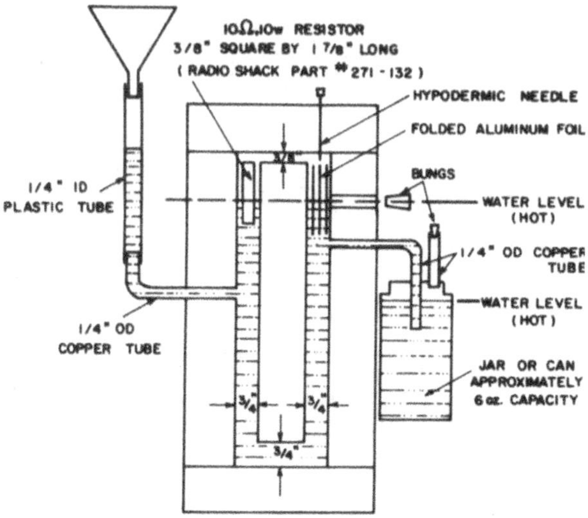

Figure 10.21 Wooden machine.

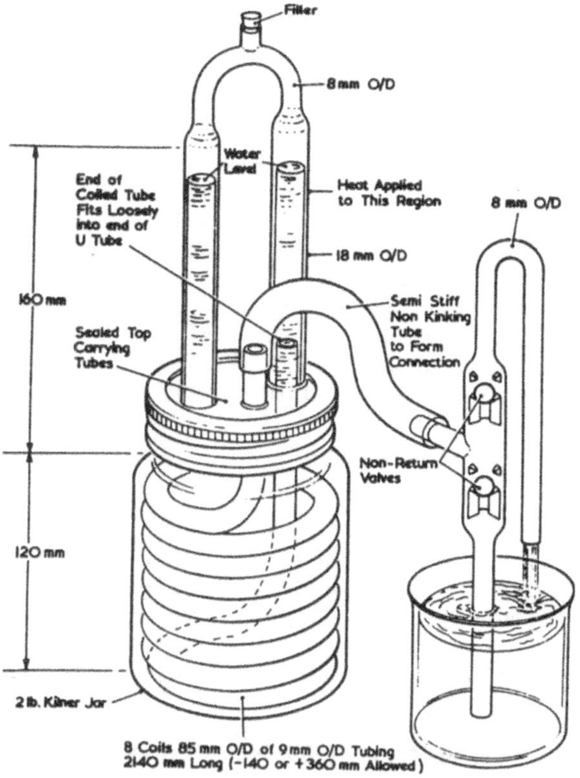

Figure 10.22 Fruit jar machine, (copies from AERE Harwell brochure).

the pressure in the engine falls, half a cycle later, cool water from the reservoir is drawn back into the cylinder.

The hypodermic needle, sealed in place with silicone caulk, is a refinement that is not essential to the operation of the engine; it provides a slow leak that keeps the average pressure inside the machine close to atmospheric and thus makes the average level of the water in the cylinders and the tuning line equal. Without such a leak, it will be necessary to remove the level plug from time to time in order to equalize the pressure.

Simple though this machine is, there are still worthwhile and original experiments to be done with it. For example, how does the amplitude of the oscillation in the tuning column of this machine change as its length is increased? What happens if the tuning line is inserted higher up or lower down in the side of the hot cylinder? Is there enough power available to drive a small pump? Can this method of construction be scaled up?

A design for a small combined engine and pumping system, published at Harwell is based upon a fruit jar (Figure 10.22). The pump uses ball valves made from glass beads or ball bearings. The glass U-tube forming the cylinders and regenerator may be difficult to acquire without access to glass blowing facilities, and could perhaps be replaced by separate tubes for each cylinder connected by plastic or rubber tubing. This machine, which will pump 5 gallons/hr (Mosby (1978)) can be heated by the focussed light from a projector lamp or from the sun, or by a hot air blower (although an ordinary hair dryer may not give a high enough temperature.

These two examples are given to illustrate the simplicity of the liquid piston system and to encourage experimentation in the field. The best way to learn something about Fluidyne engines is to build one and operate it. And there are many questions waiting to be answered, many experiments waiting to be done, and many new ideas to be had in this rather new and undeveloped field.

FUTURE DEVELOPMENT

Figure 10.23 illustrates the much larger machine, referred to earlier as a laboratory prototype. It has a throughput of more than 450 US gallons/hr at a head of 10 ft and is, of course, much higher and more complicated than the other two machines described above.

Two areas in need of further innovation are the pumping head and the tuning line. The machines built by Harwell and Metal Box have a pumping head capability of 10-12 feet. This is appropriate for many irrigation applications, but there are large areas of the world (including many states of India) where the well depth is more typically 20 feet, and a larger pumping head than is available from

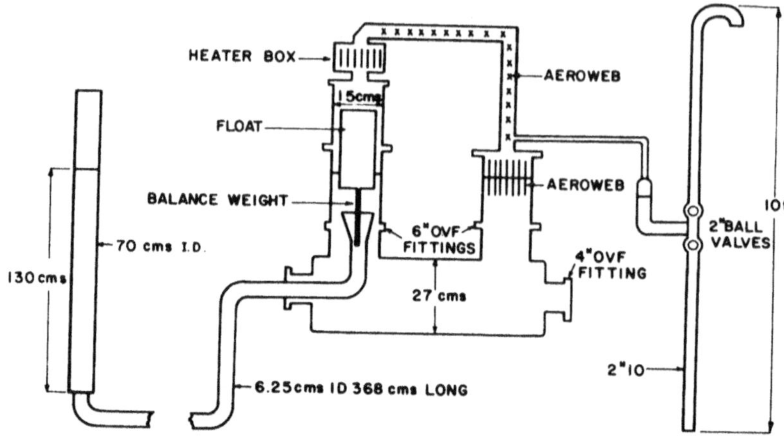

Figure 10.23 Six inch cylinder machine, (after West and Pandey (1981)).

present machines would widen the field of application. One way to achieve this increased head is by pressurizing the working fluid (West, Geisow and Pandey (1976)) and another is by using two or more pumps each gas coupled to the main Fluidyne. Pressurization would reduce the relative importance of some of the losses. Using two sets of values in a double pump arrangement (Figure 10.24) facilitates the switch from a high head configuration, with the pumps arranged one above the other, to a high throughput system, with the two pumps side by side. In both cases, the basic simplicity of the Fluidyne system is compromised and manufacturing costs are raised; there is a real need for innovative thinking in this area.

On a large machine, a tuning line designed according to equation (10.5) needs to be long and of large diameter if the flow losses are to be acceptably small. This adds considerably to the costs of the engine, and may also be a substantial source of power loss. It is possible to make the tuning line from a spiral or baffle arrangement (Figure 10.25), but this still commits a fairly large volume to this component, and may lead to even larger flow losses. New ways to carry out the tuning line function, or experimental evidence that tuning lines much shorter than resonant length are acceptable, would reduce the cost, improve the portability and perhaps increase the efficiency of large machines.

Finally, we note that entirely new applications may be found for this class of machine, applications in which the simplicity, freedom from maintenance or ability to use low grade fuel are important, and the low power density is relatively unimportant.

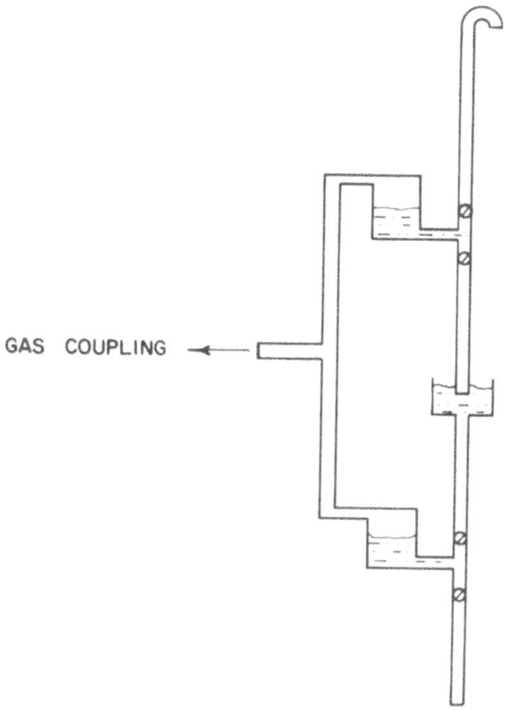

Figure 10.24 Using two valve systems in cascade to increase pumping head.

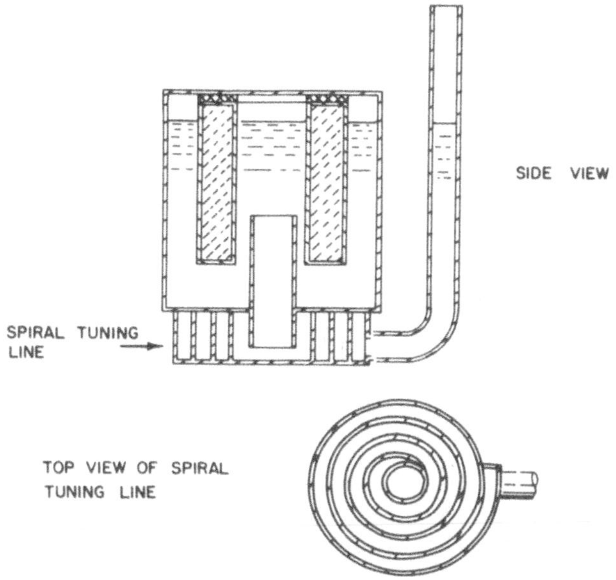

Figure 10.25 Compact low cost tuning line system applied to concentric cylinder machine, (after West, Geisow and Pandey (1977)).

NOMENCLATURE

A	cross-sectional area of displacer columns
A_s	surface area of cylinder
A_t	cross-sectional area of tuning column
f	operating frequency
F	enhancement factor for transient heat transfer loss
g	acceleration due to gravity
h	length of displacer column in common reservoir machine
k	thermal conductivity
L_d	liquid length in displacer column
L_t	liquid length in tuning column
P_m	mean pressure of working fluid
R	radius of tube
t_d	thickness of boundary layer in oscillating flow
T_m	mean temperature
V_m	mean volume of working fluid
x	displacement of liquid surface
ρ	density of liquid
ΔP	change in working fluid pressure
η	viscosity of liquid
ω	angular frequency
α	thermal diffusivity of working fluid
γ	specific heat ratio of working fluid
ΔT	amplitude of temperature variation

REFERENCES

Bell, G.C. (1979a). Passive Solar Water Pump. Independent Project Arch. 589, University of New Mexico.

Bell, G.C. (1979b). Solar Powered Liquid Piston Stirling Cycle Irrigation Pump. SAN-1894/1, April.

Breckenridge, R.W. Jr., Heuchling, T.P. and Moore, R.W. Jr. (1971). Rotary Reciprocating Refrigeration System Studies, Pt. 1, Analysis. Arthur D. Little Inc., Technical Report AFFDL-TR-71-115, Part 1, Sept.

Crandall, I.B. (1927). Theory of Vibrating Systems and Sound. Van Nostrand.

Cutler, D. and Hanke, C. (1979). Test Report of Two Fluid Piston Heat Engines. Chicago Bridge and Iron Co., CBI Research Contract R-0268.

Drzewiecki, T.M. (1979). An Initial Model for the Finite Displacement Response Characteristics of a Fluidyne Pump. HDL-TR-1868, Feb.

Elrod, H.G. (1974). The Fluidyne Heat Engine: How to Build One - How It Works. ONR London Report R-14-74, (NTIS No. AD/A-006-367), Dec.

Geisow, A.D. (1976). The Onset of Oscillations in a Lossless Fluidyne. AERE M-2840, Oct.

Gerstmann, J. and Friedman, Y. (1977). Liquid Piston Heat-Actuated Heat Pump and Methods of Operating Same. U.S. Patent No. 4, 148, 195, (Filed December 1977).

Gill, P.F. (1980). The Mathematical Modelling of a Jet-Stream Fluidyne. RNEC-SERF-F1-80, May 1980.

Goldberg, L.F., Rallis, C.J., Bell, A.J. and Urielli, I. (1977). Some Experimental Results on Laboratory Model Fluidyne Engines. Paper No. 779255, Proc. 12th IECEC, Washington, Aug.

Goldberg, L.F. (1979). A Computer Simulation and Experimental Development of Liquid Piston Stirling Cycle Engines. M.Sc. Dissertation, University of the Witwatersrand, Johannesburg, Mar.

Goldberg, L.F. and Rallis, C.F. (1979). A Prototype Liquid-Piston Free-Displacer Stirling Engine. Paper No. 799239, Proc. 14th IECEC, Boston, Aug.

Gosling, M. and Boast, D. (1976). Analog Simulation of a Fluidyne Engine. B.Sc. Project Report No. 382, University of Bath, June.

Lee, K.P., Smith, J.L. Jr. and Faulkner, H.B. (1980). Performance Loss Due to Transient Heat Transfer in the Cylinders of Stirling Engines. Paper No. 809338, Proc. 15th IECEC, Seattle, Aug.

Lewis, P.D. (1978). Operation of a Jet-Stream Feedback Fluidyne. RNEC-TR-78008, May.

Martini, W.R., Hauser, S.G. and Martini, M.W. (1977). Experimental and Computational Evaluations of Isothermalized Stirling Engines. Paper No. 779250, Proc. 12th IECEC, Boston, Aug.

Martini, W.R. (1978). Stirling Engine Design Manual. 1st Edition, NASA Report No. CR-135-382, (NTIS No. N78-23999), April.

Martini, W.R. (1983). Test on a 4 U-Tube Heat Operated Heat Pump. Submitted to 18th IECEC, Orlando, August.

Mosby, D.C. (1978). The Fluidyne Heat Engine. M.Sc. Thesis, Naval Post-Graduate School, Monterey, September.

Pandey, R.B. (1981a). Private Communication to C.D. West, April.

Pandey, R.B. (1981b). Financial Express of New Delhi, Feb. 14.

Pandey, R.B. and West, C. (1981). A Laboratory Prototype Fluidyne Water Pump. Paper No. 819787, Proc. 16th Inter. Soc. Energy Conv. Eng. Conf., Atlanta, Ga., Aug.

Park, J.R.S. and Baird, M.H.I. (1970). Transition Phenomena in an Oscillating Manometer. Can. Jnl. Chem. Eng., Vol. 48, pp. 491-495, Oct.

Rayleigh, Lord J.W.S. (1896). The Theory of Sound, Vol. 2. Second Edition, Macmillan.

Reader, G.T. (1979). The Fluidyne - A New Class of Heat Engine. Paper 19, Polytechnic Symposium on Thermodynamics and Heat Transfer, Leicester, Nov.

Reader, G.T. and Lewis, P.D. (1979b). The Fluidyne - A Water in Glass Heat Engine. M.N.S., Vol. 5, No. 4, pp. 240-245.

Singleton, J.R. (1979). The Fluid Mechanics of the Jet-Stream Fluidyne RNEC-SERG-6-79, May.

Stammers, C.W. (1979). The Operation of the Fluidyne Heat Engine at Low Differential Temperatures. Jnl. of Sound and Vibration, 63(4), pp. 507-516.

West, C.D. (1970a). Hydraulic Heat Engines. AERE-R 6522, Sept.

West, C.D., Cooke-Yarborough, E.H. and Geisow, J.C.H. (1970). Improvements in or Relating to Stirling Cycle Heat Engines. British Patent No. 1 329 567, (Filed Oct. 1970).

West, C.D. (1971). The Fluidyne Heat Engine. AERE-R 6775, May.

West, C.D. (1974a). Improvements in or Related to Stirling Cycle Heat Engines. British Patent No. 1 487 332, (Filed Nov. 1974).

West, C.D. (1974b). Improvements in or Relating to Stirling Cycle Heat Engines. British Patent No. 1 507 678, (Filed Nov. 1974).

West, C.D., Geisow, J.C.H. and Pandey, R.B. (1976). Improvements in or Relating to Stirling Cycle Heat Engines. British Patent No. 1 581 748, (Filed April 1976).

West, C.D., Geisow, J.C.H. and Pandey, R.B. (1977). Improvements in or Relating to Stirling Cycle Heat Engines. British Patent No. 1 581 749, (Filed Jan. 1977).

West, C.D. (1982a)). Liquid Piston Stirling Engines. Van Nostrand Reinhold Company, New York.

West, C.D. (1982b)). Performance Characteristics of Wet and Dry Fluidynes. Proc. 17th IECEC, Los Angeles, August.

West, C.D. (1983a). Dynamic Analysis of the Fluidyne. Submitted to 18th IECEC, Orlando, August.

West, C.D. (1983b). Stirling Engines with Controlled Evaporation of a Two-Phase Two-Component Working Fluid. Submitted to 18th IECEC, Orlando, August.

Wood, G. (1980). Lecture Notes for Stirling Engine Workshop. Sunpower Inc., 6 Byard St., Athens, Ohio 45701, October.

SUBJECT INDEX

Adiabatic processes - 113
Aerodynamic friction - 113, 118
Aerojet-General heart system - 199
Air as working Fluid - 217
Air compressor conversion - 217
Air preheater - 12
Alpha West Analysis program - 114
Alternator losses - 182, 185
Amplitude - 111
Analog computer - 110
Area ratio - 225
Articulated diaphragms - 173
Artificial heart engines - 197-215

Balance - 31
Balanced compounded Stirling hydraulic engine - 253
Bang-bang operation - 30, 204
Basic engine power calculation - 114
Basic heat input - 115
Beale number - 102-104
Benson engine - 35
Bounce space - 23, 223
Bourne/Bush engine - 202, 206
Brake efficiency - 106
British Genius Exhibition - 188
Bucket brigade loss - 116

CM-3 80K cryocooler - 227
Carnot cycle - 5, 105
Cells - 122
Coal-fired Stirling engines - 216
Cold finger - 222
Cold tip - 222
Compression space - 3
Constrained simulation - 124
Control volumes - 122
Convective spoilers - 118, 174
Cost - 12
Cryocooler, definition - 222
Cryopumping - 222
Cryo-surgery - 222
Cup seals - 142

Dashpot load - 135
Dead space - 115, 118
Degrees of freedom - 51
Design guidelines for Ringboms - 158-159

Diaphragm bulge - 170
Diaphragm engines - 47, 209
Diaphragm hub diameter - optimum - 171
Diaphragms as pistons - 168, 207
Diesel conversion to Ringbom - 160
Diesel engine conversion - 217
Diesel engines - 11
Discontinuous piston motion - 9
Displacer - 14
Displacer annulus pumping loss - 117
Displacer kinetic energy loss - 156
Dry Fluidynes - 248
Duplex Stirling heat pump - 129, 140, 217

Efficiency - 11
Electrical heating of engines - 133
Ericsson engines - 1, 7, 200
Expansion space - 3
Experience factor - 101, 103, 106, 112
Explosion hazard - 217

Family tree of fpse - 29
Fatigue failure - 168
Fatigue threshold stress - 168
Ferroxdure 300 magnet material - 227
Finkelstein adiabatic analysis - 113
Finkelstein nodal analysis - 121
First order analyses of Ringboms - 106, 150-151
Fluid friction - 123, 182
Fluidized bed combustor - 216
Fluidyne engine - 232
Fluidyne refrigerator - 36
Franchot engines - 38
Free cylinder engines - 20, 129
Free cylinder water pumps - 139
Free-piston cryocoolers - 222-234
Frequency - 69, 104, 111
Fruit jar Fluidyne - 255

Gas conduction - 184
Gas coupling in Fluidynes - 247
Gas-fired heat pump - 218
Gas leak rate - 88
Gas lubrication - 20, 90-94, 166
Gas spring hysteresis - 57-58
Gas spring types - 120
Gedeon's third order analysis - 123

Hard anodized aluminum - 89
Harmonic operation - 30, 53, 108
Hot air engines - 1
Hot gas engines - 1

Hybrid Stirling engines - 41-45
Hydraulic output - 143
Hydraulic output of Stirling engines - 210
Hydrostatic drive - 35
Hysteresis loss - 119, 138

Ideal Stirling cycle - 3-5, 101
Inertia freon compressor - 132
Inertia water pump - 129
Integral-Stirling cryocooler - 223-227
Intersociety Energy Conversion Engineering Conference - 197
Isothermal analysis - 107
Isothermalized spaces - 114
Isothermalizer in Fluidynes - 255

Large Stirling cryocoolers - 231
Large Stirling engines - 216-221
Life of engine - 168
Linear alternator - 129, 136
Linearization - 110
Liquid feedback in Fluidynes - 242
Liquid piston Stirling cryocooler - 232-233
Liquid piston Stirling engines - 235-258

M.C. 80 cryocooler - 226-227
Martini driven displacer - 161, 217
Martini multiple cylinder refrigerating machine - 39
Martini second order analysis - 114
Mass ratio - 225
Metallurgical limit - 11
Method of characteristics - 123
Military applications - 222
Model 10 engine - 129, 136
Model 100 engine - 136
Model Stirling engines - 197
Moving coil alternator - 176
Moving magnet alternator - 176
Multicylinder Fluidynes - 235, 253

Natural frequency - 53
Natural gas liquifiers - 141, 219-221
Net heat input - 115
Net power output - 115
Nodes - 122

Optimization - 124
Orowan's rule - 141
Oscillating flow in Fluidynes - 249
Overdriven mode operation - 147, 148
Overdriven mode theorem - 155
Overspeeding - 150

Pendulum engines - 45-47
Phase angle - 111
Phasor - 35
Piston - 13-14
Piston centering - 84-86
Piston rings - 142
Pneumatic cushion - 146
Poiseuille laws - 249
Power calculation - 113
Pumping application of Fluidynes - 245
Pumping power - 124, 142, 180

RE-1000 engine - 135
Radiation heat transfer - 118
Radiation shields - 118, 174
Radio-isotopes heat source - 185, 188, 199
Rankine heat pump system - 132
Regenerator - 3
Regenerator losses - 118, 184
Relative efficiency - 8, 106
Ringbom engines - 217
Ringbom Stirling engine - 139, 143, 145
Rocking beam feedback - 239
Rulon - 231

Samarium/Cobalt magnets - 227
Scaling rules - 142, 216
Schmidt assumptions - 106
Seal leakage loss - 118
Second order analyses - 112
Self-starting - 167, 208
Separate cylinder arrangement - 15
Shuttle heat transfer - 116
Side force - 20
Siemens engines - 38-39
Single-cylinder Ringbom - 161
Small Stirling engines - 197
Sodium heat pipes - 216
Spring stiffness - 96
Split-Stirling cryocooler - 15, 79, 223, 227-231
SQUIDS - 233
Stable operation of Ringboms - 149
Stainless steel - 94
Starting - 129
Static friction - 167
Stirling cryocooler - 36
Stirling cycle heat pump - 6
Stirling cycle pressure generator - 6
Stirling cycle refrigerator - 5
Stirling Engine Newsletter - 197
Stress limits for diaphrams - 169

Sunpower third order analysis - 114
Super Demo engine - 136

Temperature difference in heat exchangers - 182, 183
Thermal conduction - 116, 182
Thermal loading of heat exchangers - 186
Thermocompressor - 7, 199-204, 210
Thermo-mechanical generator - 173
Third order analyses - 121
Three-mass system - 32
Tidal regenerator engine - 211
Top end tuning - 139, 158
Transient heat transfer loss - 251
Transit tune of displacer - 150
Tuning of liquid columns - 236, 245
Two-mass system - 32

UK National Data Buoy - 188, 190

Viscous losses in gap - 103
Vuilleumier cycle - 49

Water lubrication - 160, 217
Water pumping - 129
Wet Fluidynes - 248
Wooden Fluidyne - 253
Workspace spring effect - 104, 105

NAME INDEX

AGA Navigation Aids Ltd. - 191
Agbi - 27
American Gas Association - 131
Argonne National Laboratory - 121
Atomic Energy Research Establishment at Harwell - 83, 235

Beale, William - 27, 50, 128
Bush, Vanevar - 7

Chellis, Fred - 231
Chicago Bridge and Iron Co. - 253
Cooke-Yarborough, E.H. - 47, 83, 166, 235
Cryogenic Technology Inc. - 227

Eaton Corp. - 133
Ericsson, John - 16

Gedeon, D. - 123, 124, 134, 216
General Electric - 133
General Motors - 16

Higa, Walter - 231
Horn, Stuart - 231
Hughes Aircraft Co. - 227

Jet Propulsion Laboratory - 134, 231
Joint Center for Graduate Study - 206

Magnavox Inc. - 226
Martin-Marietta - 227
Martini, W. - 197, 235
Martini Engineering - 197
McDonnel Douglas Co. - 206
Mechanical Technology Inc. - 133
Metal Box Co. - 255
Mitchell, L. - 110

NASA Lewis Research Center - 135, 209
National Energy Software Center - 121
National Institutes of Health - 198

Oak Ridge National Laboratory - 235
Ormat Turbines - 133

Philips Research Laboratories - 2, 226
Postle, Davy - 44

Rainbow, Horace - 45
Ringbom, Ossian - 44, 145

Siemens, Sir Charles - 18
Stirling, James - 2
Stirling, Robert - 2
Sunpower Inc. - 28, 83, 110, 128-144

Thermo-Electron Corp. - 211
Trinity Lighthouse Service - 191

van Weenan, Ir. - 18

West, Colin - 232, 235
Westware Co. - 114
Wood - 86, 100, 112

GPSR Compliance

The European Union's (EU) General Product Safety Regulation (GPSR) is a set of rules that requires consumer products to be safe and our obligations to ensure this.

If you have any concerns about our products, you can contact us on ProductSafety@springernature.com

In case Publisher is established outside the EU, the EU authorized representative is:

Springer Nature Customer Service Center GmbH
Europaplatz 3
69115 Heidelberg, Germany

Batch number: 10761697

Printed by Printforce, the Netherlands